The Speed of Light

Reviewing the History of "c"

By Mark F. Dennis

Illustrations by

Mark Dennis

Cover Image of the Einstein Cross courtesy of NASA and
NASAIMAGES.ORG

Sombrero Galaxy

M104

NGC 4594

"From all these results, if they were to be confirmed, would issue a wholly new mechanics which would be characterized above all by this fact, that there could be no velocity greater than that of light, any more than a temperature below that of absolute zero. For an observer, participating himself in a motion of translation of which he has no suspicion, no apparent velocity could surpass that of light, and this would be a contradiction, unless one recalls the fact that this observer does not use the same sort of timepiece as that used by a stationary observer, but rather a watch giving the 'local time'"

(Poincaré, 1904, p. 253)

Table of Contents

Table of Figures

The Speed of Light

of Light

Reviewing the History of "c"

Preface

For millennia, mankind has been looking up at the stars wondered about the universe. Where did we come from and what are we made of? One of man's greatest curiosities was about light, its composition and how it propagated through empty space, even back to the 300 BCE's, in the days of Plato, Aristotle and Socrates. It wasn't until Galileo in the early 1600's that the question of velocity of light was truly tested with the use of covered lanterns. Not much later in the late 1600's, Ole Rømer determined through the study of Jupiter's moons that the speed of light was be 299,792 km/sec or 186,290 miles/sec, becoming the cornerstone of quantum physics, the study of atomic matter.

Various theories of what light consisted of, slowly developed, including René Descartes wave theory in the early 1600's and Sir Isaac Newton's particle theory in the early 1700's. Newton's particle theory however did not stand very long as Thomas Young in the late 1700's used the "Double Slit" experiment to show that light created interference patterns, proving that light was composed of waves. Studies soon showed that light was made of both longitudinal waves and transverse waves complicating the aether theory of the day.

The aether or medium, through which light moved, at this time, was thought to be dragged along with the earth through space. In the late 1800's Albert Michelson and Edward Morley went on to disprove the existence of the aether with their interferometer experiment. This experiment confirmed to scientists of that day that there was no aether; realigning modern science with Newton's theory that light was a particle, the photon. Not only did these tests show that there was no aether, they had also discovered what is now called "Law of the Constancy of the Speed of Light." This measurement of the speed of light is the same no matter what the observer's velocity is or what velocity the emitter is; either moving towards or away from each other expressed as "c". Today this measurement stands at 299,853 km/sec. or 186,328 miles/sec, very close to Rømer's calculation hundreds of years ago.

History of "c"

The Luminiferous Aether

Before the year 1900, most scientists believed that light was propagated through space in what was called the "Luminiferous Aether". They believed that like sound through air, light must travel through something. The belief was that the light bearing "aether" was a fine gas-like substance made of the smallest of particles.

350 B.C.E - Aristotle Ancient Greeks believed that all things were made of combinations of the four basic substances, fire, water, earth, and air. Among the ancient Greeks were Greek philosophers **Plato** (427-347 BCE) and **Aristotle** (384 BC - 322 BCE), born in the height of ancient Greek civilization. Plato, a student of **Socrates** (469 BC–399 BC), was a philosopher and mathematician who laid the foundations of modern philosophy. Aristotle is believed to be the first person to attempt physics and even gave physics its name – which is the "study of matter" and its motion. Aristotle is one of the central figures of the era and was a student of Plato (Bord, 2000, p. 33).

Aristotle and Plato before him believed that in addition to the four basic elements there was a fifth element which Aristotle would call the aether, meaning "blazing" (Aristotle, De Caelo, Book I, Chapter 2). This aether would be the main constituent of the celestial bodies (Asimov, Isaac, p. 6). It would be a substance rigid enough to transmit light at incredibly high speeds while offering no resistance to the movement of the galaxies, stars and planets throughout the universe (Barbour, Julian B., p. 128). This was the belief for thousands of years until around the turn of the 20th century when new scientific discoveries came to light, no pun intended, and changed things. Throughout this discussion we will at times call what is sometimes referred to as the aether, a "*field*" or a "*medium*".

1638 - Galileo In Aristotle's time it was believed that the speed of light emanating from the celestial bodies was instantaneous or infinite. Although a few thinkers throughout the centuries suggested that light might be finite, it wasn't until Galileo in 1638 that an experiment would be devised to measure the speed of light.

Galileo (1564-1642) lived in Italy during the Renaissance, one of the most productive eras in human history. He was born the same year as Shakespeare and born the same year that Michelangelo died. Galileo gave science the gift of many discoveries in both mechanics and astronomy, such as the sun centered heliocentric model of or solar system, which led to him being placed on trial by the Roman Inquisition. Galileo also discovered the law of falling bodies. Aristotle, hundreds of years earlier had reasoned, erroneously, that the velocity of falling objects were dependent on their weights. Galileo would later discover that the velocity of two objects of different weights fell at the same rate. Legend has it that this discovery was made from the Leaning Tower of Pisa (Bord, 2000, p. 35).

In an effort to understand the velocity of light, Galileo proposed that an experiment over several miles using lanterns, telescopes and shutters be used. By measuring the time it took Galileo to see his assistant's light and knowing the distance of the lamps, Galileo reasoned that he could determine the speed of light.

Galileo and an assistant with covered lanterns went to the tops of two hills about one mile apart. The idea was that Galileo first would uncover one lantern, and then as soon as the assistant saw the light of the first lantern, the assistant would uncover his own lantern. The time would be measured from when the first lantern is uncovered until the light was seen from the second lantern. The time calculated was then to be divided by twice the distance between the hilltops, which was a total of two miles. The speed of light should be, **Velocity = 2 miles/ number of seconds.** No noticeable difference was noted during this experiment, thus proving to Galileo that the speed of light was infinite or that light travels at least 10 times faster than sound – far beyond perception of the thinkers of that time (Bartusiak, Marcia, p. 24). After Galileo's death this experiment was later performed again at the Accademia del Cimento of Florence in 1667 with no delay noted.

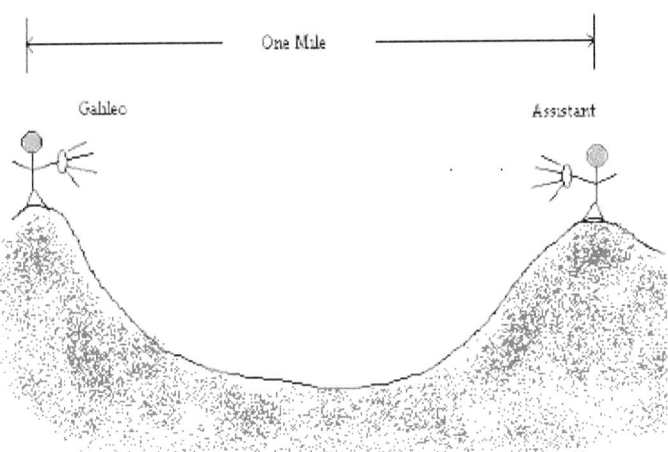

Lantern Experiment

1676 - Ole Rømer (1644-1710) was born two years after Galileo's death and was one of the first scientists to discover the speed of light through studies of the movement of one of Jupiter's moons. Romer was a student of Italian mathematician and astronomer, Giovanni Cassini, who discovered of the Great Red Spot on Jupiter in 1665.

Jupiter takes longer to orbit the sun than the earth does (twelve earth years), so that at different times of the year, the earth in its orbit around the sun moves at a high velocity toward Jupiter. At other times of the year, the earth moves away from Jupiter at a high rate of speed. Rømer would build on Cassini's observations during his studies of Jupiter, that the times between Jupiter's moon Io's eclipses got shorter as Earth approached Jupiter, and longer as Earth moved farther away. In 1676, after studying about 140 eclipses of Io, Rømer calculated that the speed of light must be 299,792 km/sec or 186,290 miles/sec (Born, Max p. 91). We will examine this later.

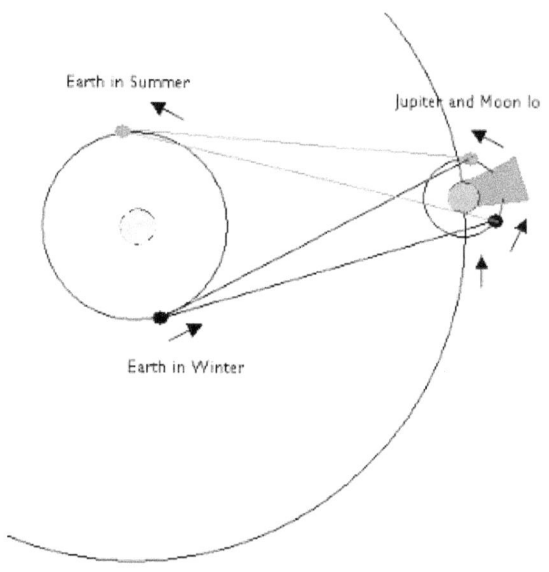

Earth in Summer

Jupiter and Moon Io

Earth in Winter

Jupiter's Moon Io

Star Dust

Now that the speed of light was known and considered finite, scientists began to question how light, magnetism, and gravity could traverse empty space. They wondered what light traveled through that limited its velocity. Empty space would have to be filled with a very fine imponderable substance they called the aether that would be the carrier or the medium in which light would travel, manifesting itself as the fields in which energy moved through empty space (Born, Max p. 86).

Two theories of light propagation existed during this era, the **corpuscular theory** or **particle theory** and the **undulation theory**, otherwise known as **wave theory**. French philosopher **René Descartes** (1596-1650) revived the idea of old that the aether was the carrier of light as well as developing the law of refraction, also known as Descartes' law. **Sir Isaac Newton** (1643-1726) however was regarded as the author of the corpuscular theory, suggesting that light was something that moved out and away from bodies like "hot ejected particles" (Born, Max p. 88).

Descartes' Luminiferous Aether would be an **elastic medium** made up of the smallest of particles, the God Particle or stardust if you will, permeating everything. A substance that oscillates, setting in motion other particles in waves, moving its energy through "empty" space. Diffraction, refraction and reflection of light all seemed to support this hypothesis that light moved in waves. This aether however **could not impede the motions or cause resistance to the planets or stars and yet must be rigid enough to transmit light** (Barbour, Julian B., p. 128). The earth therefore would move through this aether, possibly causing an **aether wind** to be detected across the earth's surface on its journey around the sun.

With Descartes undulation or wave theory, this aether would have to be considered an **elastic solid** due to the fact that light is transmitted as a transverse wave. Transverse waves cannot propagate through the fluidic motions of liquids and gases such as ocean waves and sound wave through air; this is because there is no resistance to lateral displacements of particles in these waves. Solid bodies propagate waves that are both longitudinally and transversely; fore and aft, left and right and up and down as evident with the polarization of light (Isaacson, p. 115).

According to the wave theory, light would theoretically travel through the aether in waves, very much like sound through air. A wave is defined as a traveling disturbance of coordinated vibrations that transmit energy through a substance **without net movement of that substance.**

There are two distinct types of waves; **longitudinal waves** and **transverse waves.** Both waves can travel in elastic solid mediums, however, waves traveling through fluids such as liquids and gases are always longitudinal waves. Longitudinal waves are waves in which oscillations are along the direction of wave travel such as in sound waves where there is a back and forth motion of these particles. The absence of rigid bonds between particles in a fluid prevents traverse waves from traveling through them. A transverse wave is a wave in which oscillations are perpendicular to the direction of wave travel such as a wave on a rope tied to a doorknob, electromagnetic waves and seismic waves traveling through the ground, calling for a somewhat rigid medium in order to conduct their energy (Tipler, p. 465).

Transverse Wave

One way to view a longitudinal wave is to string single particles along in a line, somehow tied together with spring like forces. As one particle is forced away from another, an effect is felt on the particles to its left and to its right. One is pulled along for the short distance it moves and the other is pushed. A chain event takes place as one particle pulls on another and one pushes on the other down the line in one direction longitudinally, also known as a compression wave.

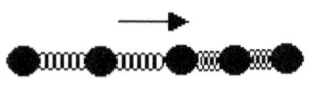

Longitudinal Wave

6

Now, visualize if you will, not only fore and aft but up and down and left and right motions in which all particles effect each other in all directions

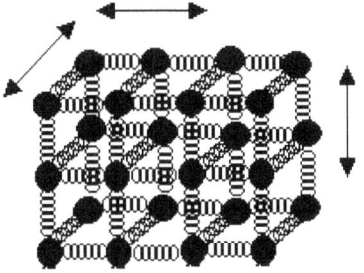

Transverse Wave

Polarization is the orientation of these transverse waves in light waves. Transverse light waves not only are propagated in a certain direction at the speed of light, they also oscillate perpendicularly to the direction of propagation, vertically and horizontally at the speed of light (Bord, p. 326). The fact that light waves show properties of polarization reveals that light waves are indeed transverse waves. Therefore if light is propagated through a medium, and transverse waves can only be transmitted through solids, then **this medium that light moves through must then be an elastic solid,** somewhat like an "elastic jelly" (Psillo, p. 131).

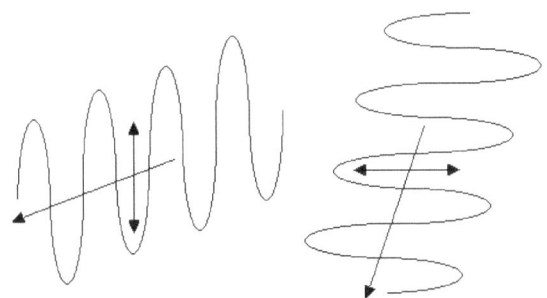

Horizontally and Vertically Polarized Transverse Waves

Light however seemed to also show signs that it was also a particle. Newton believed that through his studies with prisms, that light was made up of particles that moved at different velocities. Blue light particles for example, moved faster than red light particles, accounting for the different angles or refraction when passed through a prism. The fact that light was a particle also was manifested to him in that light traveled in a straight line. But Newton also recognized the phenomenon of diffraction and interference that could only be best explained when light was considered a wave, leading to a confusing view of **duality,** light being both waves and particles during the late 17th century (Bord, pr. 364).

Newton's particle theory however did not stand very long. **Thomas Young** (1773-1829), a brilliant Englishman, showed through the "**Double Slit**" or "**Two-Slit**" experiment, that light generated **interference patterns** and that light waves of different colors had different wavelengths. Through studies with polarized light, he discovered that light was a transverse wave, which supported the **undulation** or **wave theory** (Brown, p. 87).

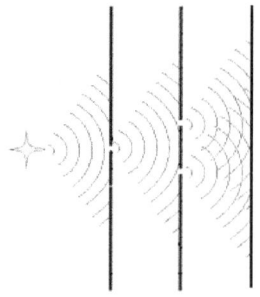

Double Slit Experiment

Star Aberration

During the eighteenth century, the Copernican heliocentric view of the universe had generally been accepted although there was very little empirical data to support it (Born, Max p. 10).   It was for this that Galileo in 1633 had risked martyrdom for the sake of what he believed to be truth.

In December 1725, one year before Newton's death, English astronomer **James Bradley** (1693-1762) began a prolonged observation of Gamma Draconis (also known as Eltanin), a star in the Draco constellation. This constellation is a group of circumpolar stars that remain in the night sky all year around and which passes directly overhead nightly where he lived in London. The study was an effort to the view the parallax of this star, using triangulation to determine its distance from the earth.

Throughout his observation viewing **Gamma Draconis** through a telescope fixed to the chimney on the roof, Bradley found that he had to continuously adjust his telescope to a different position almost nightly. Over a six months period, the position of the star seemed to move one direction then the other from its central position with a maximum variation of approximately 20.5 arc seconds. Unlike the circular motion expected due to the earth's tilt on its axis, the star seemed to move in an elliptical pattern. Bradley discovered that this angle of "aberration" depended on the ratio between the orbital velocity of the earth and the speed of light. Knowing the velocity of the earth, 30 km/sec, Bradley was able to determine the velocity of light very close to the calculations of Ole Rømer back in 1676 (Van der Kamp, Walter p. 27). This interpretation seemed to work well with either the corpuscular theory or the wave theory.

Let's look at this in another way. A tall box is designed that is totally enclosed except for a pinhole in the top that allows sunlight to shine on a small point on the floor. The box being on earth is moving at 30 km/second in its orbit around the sun. As the light shines through to the bottom of the box, the box moves, causing the light to strike off-center of the bottom of the box.

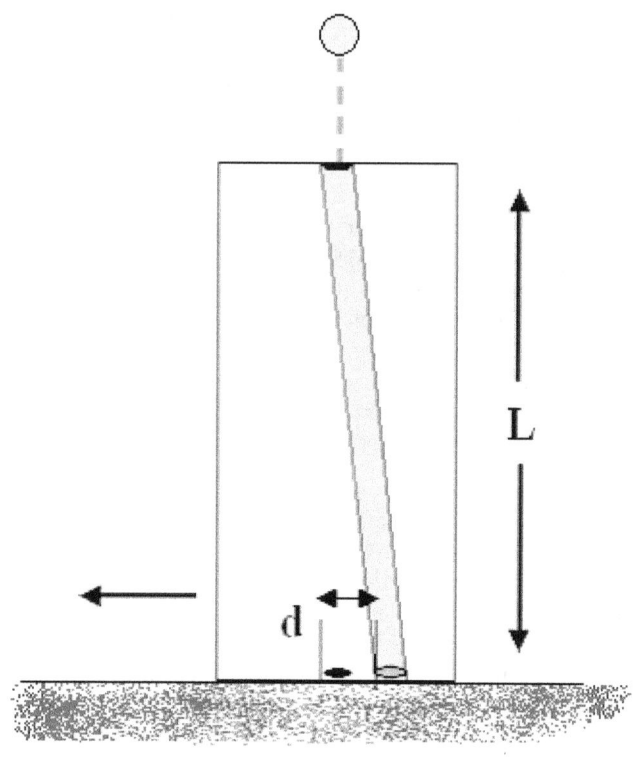

Tall Box Analogy

With the height of the box = L and the distance off center = d, we can use the formula d = v x l / c. The speed at which the earth goes around the sun is: v = 30 km/s, therefore the speed of light will be c = 299,792,458 m/s. (Born, Max p. 95). This ends up being a very small numerical value; however, through the multiplication effects of a telescope's lenses and through fine adjustments, this amounts to 20.5 arc seconds in displacement of the telescopes angle in the direction of the earth's motion.

$$\frac{d}{l} = \frac{v}{c}$$

Star Aberration

Star aberration or **stellar aberration** is caused by a slight displacement of a stars visible position when viewed through a telescope; this change in position is independent of the stars distance, unlike stellar parallax. One way to visualize this is by considering what happens to rain as it falls down a vertical stovepipe carried by someone walking in the rain. If the person is stationary, the rain falls straight through to the bottom of the pipe. If the person carrying the pipe begins to walk, the rain falling through the stovepipe will begin to hit the inside wall of the pipe. The pipe would have to be held at an angle in order for the rain to fall straight through. Furthermore, the faster the person walks, the more tilted the pipe must be held to allow the rain to continue falling through. Another way to view this would be a person holding an umbrella and walking. The faster one walks; the more tilt is needed on the umbrella. This stellar aberration also can be observed only with the movement of the observer and completely disappears if the observer is stationary and the light source is moving. A way to understand this is for example; a cloud that is moving overhead does not cause the rain to fall at an angle when there is no wind – the stove pipe would still have to be held vertically.

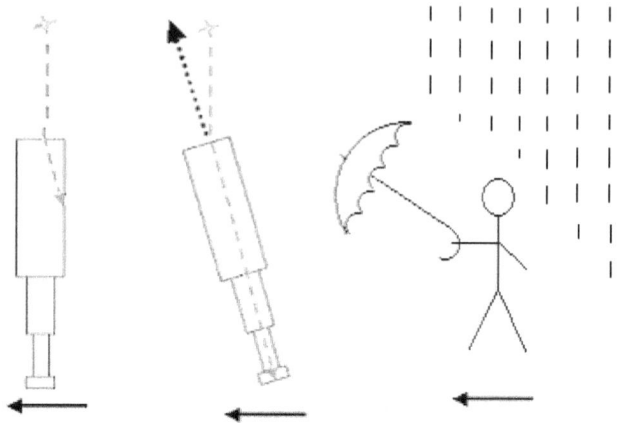

Stellar Aberration Displacement

A star viewed on the earth's ecliptic plane will vary in its position in a straight line. As an observer tilts the telescope north towards circumpolar stars, these changes in apparent position of the stars over 6 months turn into ellipses and then circles directly over the poles. The sun has a truly special case as its apparent position, staying at a constant 20.49" arc seconds. This was the empirical proof needed that the earth was in motion around the sun and that the Copernican heliocentric view of the universe was correct.

Stars on the earth's ecliptic plane can only be viewed for part of the year, becoming visible at dusk, rising at an angle in the night sky due to the earth's tilt and fade from view at dawn. These stars will change in position from night to night, first noticeable at sunrise in the east, moving towards the west a little every night until they are no longer visible being below the western horizon; these stars are then in the daylight sky and cannot be viewed. The **apparent position** of these stars when they first are visible does not shift, then as the months go by and the star rises in **Right Ascension** (RA) or (**α**), east to west, its apparent position moves in a straight line until it is most noticeable when the earth's rotation is 90 degrees to the stars light. As the star moves into **Declination** (DEC) or (**δ**) in the west, this apparent position is more noticeable and the star seems to accelerate back to its true position.

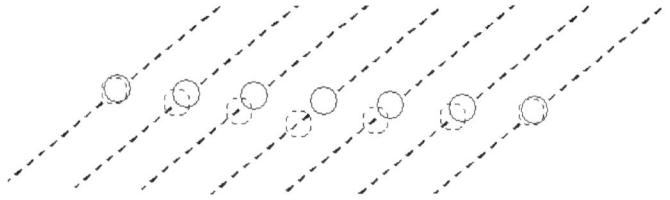

Apparent Position (ecliptic plane)

Stars in the Polar Regions are called circumpolar stars and can be viewed all year in their hemispheres. Northern circumpolar stars rotate around the North Star, Polaris, and can be seen any where between the equator and the artic circle all year. Below the equator, Polaris is not visible, dropping below the northern horizon; above the artic circle, these stars can only be viewed during the winter, as the sun does not set in the summer. These stars should rotate in a circular pattern during the year due to the tilt of the earth's axis but move in an elliptical motion due to stellar aberration; Polaris is the only star that remains in a circular pattern. The further south the star is, the more elliptical its motion becomes. These circumpolar stars shift in their apparent position in one direction in the winter months and in the other direction during the summer months, depending on their position in the night sky. This apparent shift in position is most noticeable when the earth's orbit is at 90 degrees to the starlight.

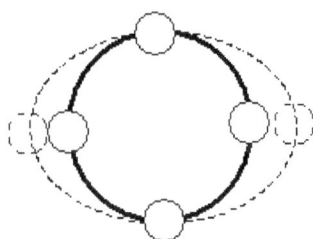

Apparent Position (circumpolar stars)

The star being viewed is so far off that the angle of starlight reaching the earth is the same throughout the year. Most stars can only be viewed for a part of the year due to their position. Stars lower on the horizon move into the daylight side of the earth for six months of the year or so and cannot be viewed; the sun will even eclipses some of these stars. When these stars first appear, they become visible just before dawn (position A) and starlight hits the earth parallel to its movement. As they continue orbiting the sun, the angle at which light from the stars hits the earth is at 90 degrees to the earths orbit (position C). After six months or so the star fades back into daylight becoming last visible at sunset where the starlight hitting the earth is again parallel to its movement.

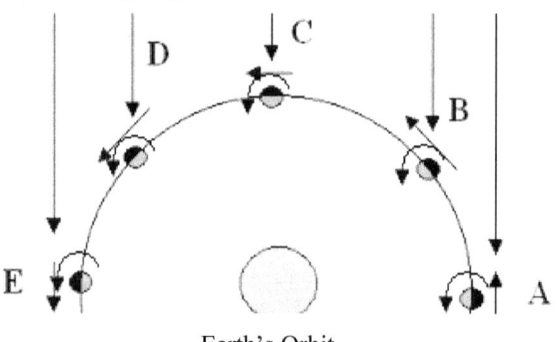

Earth's Orbit

François Arago (1786-1853) a French mathematician and physicist, assumed along with other scientists of that day, that the speed of light from a large star would be slower than light from a small star due to its massive gravitational pull on the particles of light emitted from it. In 1810, Arago, in appreciation of Newton's **refraction theory**, came up with a simple idea to detect various light velocities from stars. Since the refraction angle of light will be different for particles of light moving at different speeds, a prism mounted on a telescope when focused on different stars should be able to differentiate slow light from fast light. Slow light from large stars would refract at sharper angles than fast light from smaller stars. Stars moving towards the earth and stars moving away from earth should have different speeds of light coming from them and should have different **refraction indexes**. This experiment failed to show any deviations, indicating that light coming from all stars was exactly the same speed through the prism – even the various colors.

Arago States:

> *"By examining the receding tables attentively, one finds that the rays of all stars are prone to the same deviations...*

> *This result seems to be, with the first aspect, in manifest contradiction with the Newtonian theory of the refraction, since a real inequality in the speed of the rays however does not cause any inequality in the deviations, which they test."*

(Arago, p. 562-563)

Augustin Fresnel, (1788-1827) a French physicist, in September of 1818 responds to Arago's discovery in a letter:

> *"By your fine experiments on the light from stars, you have demonstrated that the movement of the terrestrial globe has no noticeable influence on the refraction of rays emanating from these stars..."*

> *"You have inspired me to investigate whether the result of these observations could be more easily reconciled with the system which takes light as consisting of vibrations in a universal fluid. It is all the more necessary to explain these results using this theory, since it must apply equally to terrestrial objects; for the speed with which the waves are propagated is independent of the movement of the bodies from which they emanate."*

(Fresnel, p. 57–66)

Throughout the 1800's more studies were performed in an effort to find the aether and prove that the speed of light was finite. In **1849 - Armand Hippolyte Louis Fizeau** (1819-1896) (who had predicted red shift) almost 200 years after Rømer in 1848 used a series of mirrors and half-silvered mirrors over a distance of 8.6 Km (5.36 miles) to determine light speed. Using a **spinning toothed wheel** as a timing device through an aperture, Fizeau calculated the speed of light at 313,300 Km/sec or 194,684 miles/sec (Born, Max, p. 96). Other experiments such as those with the *interferometer* and experiments involving light wave interference phenomena confirmed this.

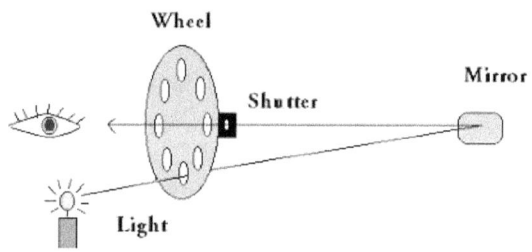

Fizeau's Toothed Wheel

1850 - Jean Bernard Léon Foucault, (1819-1868) improved upon this method bouncing light off a spinning mirror to a stationary mirror. When the light returned to the rotating mirror it reflected at a slightly different angle, which was used to determine the lights velocity improving accuracy by about 1,000 miles per second. Using the interferometer, **Albert Michelson** refined this experiment determining that light speed was 299,796 Km/sec or 186,293 miles/sec, within 3 miles per second of Ole Rømer's measurement in 1676 almost 200 year earlier (Born, Max, p. 103).

Foulcault's studies had shown that the velocity of light in water was much slower than that of light moving through the air. Bradley had suggested that since light traveled slower in water when compared to air that a **telescope filled with water** should increase the stellar aberration. In 1871, more than one hundred years later, **Sir George Biddell Airy** (1801-1892) demonstrated that stellar aberration occurs even when a telescope is filled with water and that there was no change in the aberration angle. This null result suggested a lack of universal ether (Van der Kamp, Walter p. 30).

In **1887** - **Albert Michelson** (1852-1931) and **Edward Morley** (1838-1923) set out to prove the existence of the aether using an **interferometer.** It was assumed by the experts of the day that the aether was a **rigid substance**, which was stationary in space and was the *medium* in which light traveled. It was believed that as the stars and planets traveled through space, they passed through this substance creating an "**aether wind**" that should be detectable with some sort of device. Michelson and Morley set up an interferometer (a devise that utilized mirrors and detectors to measure the speed of light) in such a way that as the earth passed through the aether wind in space, distortion or interference fringes would develop on the detector. The theory was that the speed of light would be different in each of the arms of the detector, one pointing into the aether wind (the direction the earth was moving in space) and the other perpendicular to the assumed aether wind. **No wind was detected**. (Ditchburn, R.W., Pg 315). We will also examine this a little later.

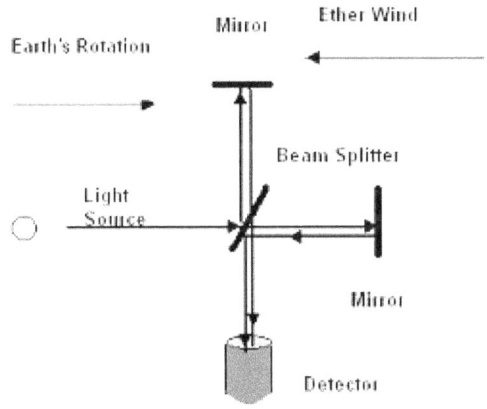

Michelson's Interferometer

The famed Michelson – Morley Experiment in 1887 with the *interferometer*, and other experiments involving light wave interference phenomena confirmed to scientists of that day that there was no aether. Not only did they seem to show that there was no aether, they had also discovered what they believed to be the "Law of the Constancy of the Speed of Light." This constant is expressed as "c" and is measured at 299,853 km/sec. or 186,328 miles/sec. **This measurement of the speed of light in air would not vary and is the same velocity no matter what the observer's velocity is or what velocity of the emitter is, either moving towards or away from each other** (Wolf, Fred, p. 52).

Partial Aether Drag Theory

*"If one supposed that our globe impresses its motion on the aether in which it is enveloped, one would easily conceive why the same prism always refracts light in the same manner, whatever the side from which the light may arrive. **But it seems impossible to explain the aberration of stars by this hypothesis**: I have not been able, so far at least, to account for this phenomenon clearly except by supposing that the ether passes freely through the globe, and that the speed communicated to this subtle fluid is only a small fraction of that of the earth, not exceeding one hundredth of it, for example"*

(Fresnel, A. p. 57–66)

Fresnel now was suggesting that the aether or medium through which light traveled, must have been sticking to the earth, being dragged along. As light from the stars would enter this medium at different velocities, it would enter the earth's atmosphere at one speed. It seems however, that if the aether drag hypothesis were true, then stellar aberration could not occur because the light would be traveling in the aether, which would be moving along with the telescope (Van der Kamp, Walter p. 30). Fresnel at first saw this as contradictory but then suggested that material with low refractive indexes, such as air, must produce almost no resistance at all to stellar light allowing stellar aberration to occur. However, substances with a higher refractive index, such as water and glass, produce more resistance, thus allowing for Arago's experiment that indicated the constancy of the speed of light.

Light would therefore travel unimpeded through the air and be uninfluenced by the movement of the earth. Once light enters into glass or some other medium of higher refractive indexes, it would therefore be constrained or locked into the new medium and travel at a different (slower) velocity within that medium. The Fresnel **partial drag theory** was widely received during the last part of the nineteenth century. This both allowed for the constancy of the speed of light and for star aberration.

.The partial aether drag theory suggested that the velocity of light would be affected by a moving medium. Fresnel projected that the aether's density in a substance was different than that of the aether in empty space and depended on the refractive index (n) of the substance it was traveling through. Air would have very little resistance to the aether but substances like water and glass would have a lot of resistance. Here c = the speed of light, n = refractive index, and v = to the velocity of the medium (Ditchburn, p. 334).

$$c' = c_1 + v\left(1 - \frac{1}{n^2}\right)$$

Fresnel's Convection (Aether Drag) Coefficient

In an effort to find some variation in the stellar aberration, Sir George Biddell Airy in 1871, as noted previously, carried this idea a little further with the use of a water-filled telescope as suggested by Fresnel in his 1818 letter to Arago. Theoretically as light would travel through the water as compared with air, refraction of light in the water would affect stellar aberration in some degree. **It did not**.

The Dilemma

The speed of light was now known. However, experiments such as the famous Michelson – Morley Experiment indicated that the velocity of light was always the same, independent of the motion of the emitter or the observer. The "simple emission theory" suggests that light given off by an object should move at a velocity relative to the movement of the emitting object and/or the movement of the observer, however this was not the case. The aether drag theory tried to answer this with an aether that was being dragged along with the earth acting as a medium, causing all light from heavenly bodies to be perceived at one velocity – like sound is through air. Stellar aberration however contradicted this by proving that light was not being carried by any medium, at least not in the earth's atmosphere. Fresnel partial drag theory tried to correct this by stating that aether drag was dependent on refractive indexes and that air had little effect on light; denser object such as glass and water however would carry light at varying degrees. Airy's water filled telescope experiment in 1871 seemed to contradict this.

In 1913 while studying **Binary Stars** or **twin stars** that orbit around each other, **<u>Willem de Sitter</u>** (1872-1934) a Dutch mathematician, physicist and astronomer, suggested that stars in these systems should have an orbit that emitted light at different speeds ($c + v$ or $c - v$). The problem was that the "faster" light given off during the orbit of a star towards the earth would be able to catch up with and even overtake the "slower" light emitted from the other star in the system retreating away from the earth. Theoretically, this difference in light velocities would cause the stars light to be scrambled and out of sequence, distorting the light seen from these binary stars (Einstein, Albert, p. 18).

The fact that there is no distortion means that we can be positively sure that the speed of light coming from both stars is the same from the instant it left the stars. With the assumption that the movement of observer is equivalent to the movement of the stars, this leads us to the conclusion that the speed of light does not depend on the speed of observer or the movement of the stars; the speed of light will always be 300,000 km/sec to the observer detecting the light (Sokolov, G. p. 2).

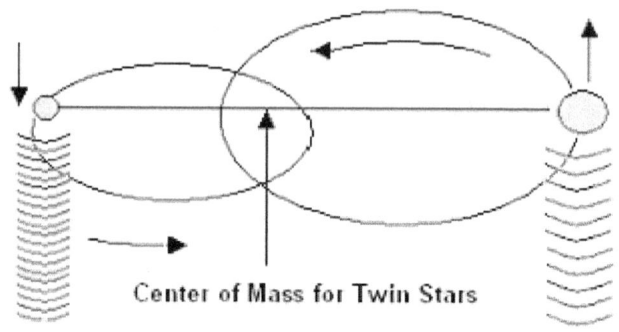

Center of Mass for Twin Stars

Doppler Effect Twin Stars

Although the light coming from these stars travels at the same velocity, the wavelength of that light is not the same. When one of the twin stars is moving towards the earth, the wavelengths compress and become shorter, towards the blue in the light spectrum, known as blue shift (similar to the effect of a whistle emitted from the caboose to an observer on the train). When one of the twin stars is moving away from the earth, the wavelengths are shifted to longer wavelengths into the red portion of the light spectrum, known as red shift (similar to the effect of a whistle emitted from the engine to an observer on a train) producing the "Doppler Effect" (Bless, p. 235).

This now creates a dilemma and raises some important questions. If light is made up of particles called photons or "hot ejected particles" as Newton would put it, and was leaving a star moving towards a detector or an observer at the speed of light, what is it that slows the photons down? If the star emitting the light being detected is moving away from the detector, what speeds up these photons? The photons must be moving at the speed of light in reference to the moving star that they came from yet be moving at the speed of light in respect to the detector or observer. How can this be?

Before the 1900's it was natural for almost everyone to believe that light must be transmitted through the material called the aether, which permeated everything. This aether would fill all of the space between the stars and the earth; otherwise light could not be visible coming from these stars. If light moved in waves, as it seemed to, just as there must be air for sound to move through, there must also be a medium for light to move through. In his book **Opticks**, published in 1704, Sir Isaac Newton wrote:

"Is not the heat of the warm room conveyed through the vacuum by the vibrations of a much subtler medium than air, which after the air was drawn out remained in the vacuum? And is not this medium the same with that medium by which light is refracted and reflected, and by whose vibrations light communicates heat to bodies, and is put into fits of easy reflection and easy transmission?"

(Newton, Sir Isaac, Book III, Part I, p.323)

We know that sound has a fixed velocity in air at a given temperature: would this also be the case with light traveling through this aether? If this aether was present throughout space, as the Earth rotates on its axis and moves in its orbit around the sun, should not this aether be possibly detected as an eerie **wind** blowing through the earth, affecting the wavelengths of light moving in specific directions. Light waves moving with the earth's rotation it seems should be compressed and light waves moving perpendicular to the earth's rotation should not be affected.

As previously discussed, the theory behind the interferometer built by Michelson and Morley was that, as the earth rotated and moved in its orbit around the sun, the aether wind that was stationary in space would flow through the detector at a specific angle. A light beam would be split with a half-silvered glass plate and be reflected off two mirrors. One beam would be pointed into the aether wind (presumed to be flowing in the opposite direction of the earth's rotation and movement through space) and the other reflected perpendicular to the aether wind. The light beamed into the aether wind would slow down compressing the waves, making them shorter. When these waves enter the detector they would be out of phase with the perpendicular beam that was not affected causing distortion on the detector. If there was no distortion there evidently was no aether wind.

The detector was build and floated on pool of liquid mercury to dampen out any noise and vibrations caused by the surroundings. After several experiments with no distortion in the detector, it was determined that there was no **aether wind** (Collins, H. p. 37). The Michelson Morley experiment seemed to confirm that light takes the same time to travel each path irrespective of the motion of the observer or detector.

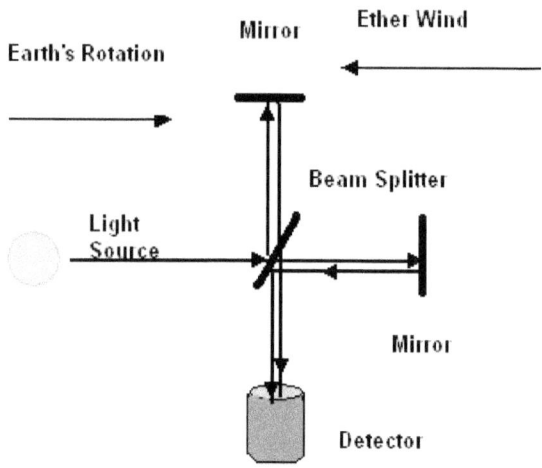

The Michelson Morley Experiment

In 1895, Dutch physicist **Hedrick Lorentz** (1853–1928) theorized that the "null" result obtained by Michelson and Morley must have been caused by an "effect of contraction" made by the aether on their apparatus and introduced the length contraction equation (Collins, H. p. 38). Where L is relativistic length and L_0 is length of object at rest. The idea here is that the aether must still exist but that high-speed movement through space distorts space enough to nullify any measurements that could detect the aether; the higher the velocity the more contraction of space. This is now known as the "Lorentz Contraction".

These discoveries at the turn of the century changed science; from this, Albert Einstein developed his "Theory of Relativity".

$$l' = l \sqrt{1 - \frac{v^2}{c^2}}$$

Lorentz Contraction Equation

Albert Einstein (1879-1955) was born in 1879 into a nonreligious Jewish family in the city of Ulm, southern Germany (Isaacson, Walter p. 11). He was born the same year of Scotsman James Clerk Maxwell's death, one of the leading scientists of the day in the field of Electricity and Magnetism, an area in which Einstein would become well versed. Apparently stubborn, young Einstein did not speak until the age of 3, waiting till he could talk in full sentences. He loved playing with puzzles, working geometry problems and enjoyed toying with magnets. Later on as a teenager, due to conflict with his teacher, Einstein would end up dropping out of high school, but was still able to enroll in a Swiss university. Although being considered "lazy" by some of the professors there, Einstein soon became well schooled in the area of physics; a field he would soon pioneer and later would develop the famous "Theory of Relativity" (Bartusiak, Marcia, p. 29-30).

The Special Theory of Relativity

Albert Einstein began to wrestle with these seemingly contradictory findings, struggling with the properties of light whether it was wave or particle and the constancy of the speed of light. After years of study, he came to a conclusion that if the velocity of light was constant, no matter what the velocity of the observer towards or away from a light source, or the velocity of the object emitting the light (as indicated by the famous Michelson – Morley Experiment in 1887) that this velocity must both change time and space.

In order to fully understand Einstein's "Theory of Relativity", one needs to understand "relative velocity". To measure the velocity of any object, you must have a means of measuring distance and a means of measuring time. Einstein visualized this in his writings with a measuring rod (yard stick, tape, what ever) and a clock (Kafatos, Menas, p. 24). Today, in the modern world we can use a baseball and a radar speed detector as an example.

If the baseball is thrown towards the stationary radar detector at 100 mph, the detector will detected the baseball at that same velocity - 100 mph. Now, let's put the radar detector in a vehicle and drive at 50 mph towards the baseball being thrown at 100 mph. This moving baseball can now be viewed as having two speeds, one speed relative to the pitcher who threw the ball at 100 mph and the speed of the baseball relative to the vehicle being detected at 150 mph; the combination of the vehicles speed of 50 mph added to the speed of the ball relative to the pitcher of 100 mph.

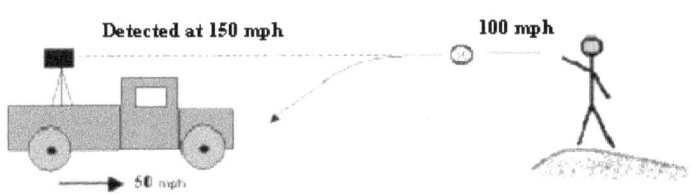

Baseball at 150 mph

On the other hand, if the detector on the vehicle is moving away from the baseball at 50 mph and the baseball itself is thrown at 100 mph relative to the pitcher, the baseball will be detected by the radar detector at 50 mph (as viewed from the moving vehicle) (Bord, 2000, p. 448).

The velocity of the baseball has to be measured relative to something, in this case, a radar speed detector moving in a vehicle or the velocity of the baseball relative to the pitcher. During the pitch, the ball is traveling at a different velocity relative to the pitcher who threw the ball and relative to the moving radar detector in the vehicle; it's all a matter of perspective. A bystander or passerby will have another relative velocity depending on their direction of movement and velocity relative to the baseball.

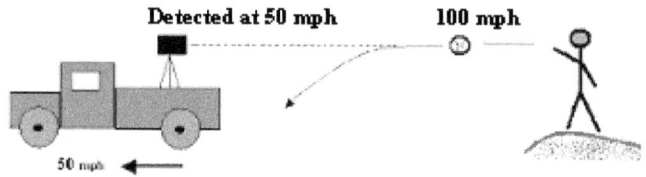

Baseball at 50 mph

The problem now is that when measuring the velocity of light, no matter what the velocity of the observer (radar gun in this case) towards or away from the light source (baseball), the measurement of the velocity of the light from this object remains the same (Born, Max, p.231-232).

We can use the earth as an example of this in much the same way as the baseball. As the earth orbits the sun, its mean orbital velocity through space is approximately 30 kilometers/sec. (18 miles/sec) or 64,800 miles per hour. As the earth in its orbit moves directly toward a star, (assuming that the star is stationary in space) the velocity of the light detected from that star, _should_ be the sum of the velocity of the earth toward the star and the speed of the light from that star, similar to the baseball and radar detector moving towards each other (simple emission theory). On the other hand, as the earth moves away from the star in its orbit around the sun, the speed of the light detected _should_ be the difference between the velocity of the earth in its orbit away from the star and the speed of the light from the star, similar to the baseball and radar detector moving away from each other (Born, Max, p. 132).

Movement of the earth towards a star should be:

EXAMPLE 1: 186,000 m/sec +18 m/sec = 186,018 m/sec

Movement of the earth away from a star should be:

EXAMPLE 2: 186,000 m/sec - 18 m/sec = 185,982 m/sec

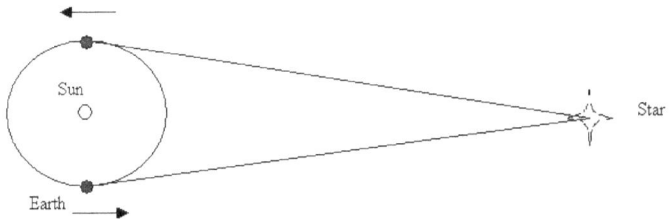

Earth and Star

But this is not the case. Sir Isaac Newton's **Classical Mechanics** and this new Theory of Relativity are now in conflict. Einstein himself said:

"Who would imagine that this simple law (the constancy of the velocity of light in a vacuum) has plunged the conscientiously thoughtful physicist into the greatest of intellectual difficulties?"

(Einstein, Albert, Pg 19).

"Law of the Constancy of the Speed of Light" was now stated:

"The velocity of the propagation of light cannot depend on the velocity of motion of the body emitting the light" (nor the motion of the observer detecting the light).

(Einstein, Albert, Pg 18).

No matter what the velocity of a detector towards or away from the star, or the velocity of the star itself towards or away from the earth (or detector), the speed of light coming from the star into the detector is always the same, 186,000 m/sec. The math doesn't seem to work – according to Einstein, space and time must be being warped.

EXAMPLE 1: 186,000 m/sec +18 m/sec = 186,000 m/sec

Earth Traveling Towards a Star

EXAMPLE 2: 186,000 m/sec - 18 m/sec = 186,000 m/sec

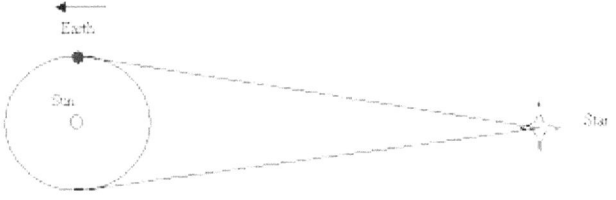

Earth Traveling Away From a Star

To add to this perspective, in addition to the earth's rotation around the sun being 30 km/sec (18 mi/sec) in reference to a star in our galaxy, the movement of the solar system rotating around the center of the Milky Way galaxy is about 250 km/sec (140 mi/s). The Milky Way itself is traveling approximately 300 km/s (185 mi/s.) in reference to other galaxies in the "Local Group". Most galaxies are spreading out away from each other at enormous velocities in an ever-expanding universe. The great **Andromeda Spiral Galaxy** however, one of the Milky Way's nearest neighbors, is actually moving toward the **Milky Way** and earth at about 50 km/s (about 30 mi/s.)

The General Theory of Relativity

Albert Einstein would then build on "Special Relativity" with the knowledge he gained in studying gravitational fields. He would go on to explain the universe and everything within it as four-dimensional "Time-Space". In three-dimensional space, the x axis would be the fore and aft, the y axis would be the side to side and the z would be the up and down with respect to the "Fixed System" reference; the fourth dimension is "time", creating Einstein's universe of Space-Time. Velocity could then be explained as the change in position over time. From this perspective, Einstein would develop the theory of the constancy of the speed of light, the theory of gravity as well as explain gravitational redshift and even time dilation.

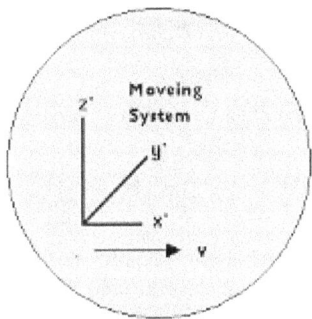

Time- Space

Gravity as Einstein would see it, exists due to "time-space curvature (Isaacson, p. 193); as if the fabric of space was being warped. Picture if you will four people holding a sheet at the four corners tightly, then someone sets down a bowling ball in the center of the sheet causing the center of the sheet to sink. A marble is then rolled onto the blanket; as it rolls, it spins around the bowling ball a few times in a tight spiral until it stops, down between the bowling ball and the blanket. From this premise, Einstein would develop his famous "**Field Equation of Gravity**":

33

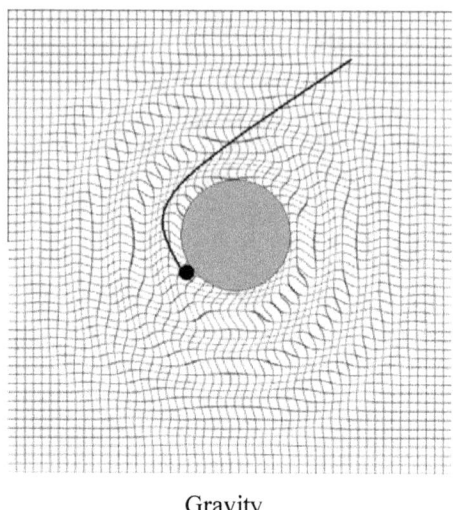

Gravity

Einstein's "Field Equation of Gravity":

$$R_{uv} - \tfrac{1}{2} g_{uv} R = T_{uv}$$

The left side of the field equation of gravity represents a geometric gravity field of space-time and the right, the distribution of mass-energy. Both sides create a balance, the more the mass-energy, the greater the warping of space-time; one affects the other and there cannot be the one without the other. Objects such as planets simply naturally follow these curves in time-space, as well does light. Einstein's new formula actually doubled his original formula in the amount of deflection that gravity would cause a ray of light to bend. Later we will see that this was the case; in 1919 when astronomer Arthur Eddington photographed a stars new apparent position during a solar eclipse, where the suns gravity actually changed the stars apparent position in the sky (Bartusiak, p.45).

We will also see that not only is motion influenced by this curvature of space, but that time also is affected (Einstein, p. 81). Recent technologies such as the Global Positioning System (GPS) must adjust for time dilation due to the earth's gravitational field and the satellites velocity (Bartusiak, Marcia, p. 55).

Relativity Re-examined

The principle of relativity is not really new. Galileo described possibly the first principle of relativity in his book, *Dialogue Concerning the Two Chief World Systems*. Here he performs a mind or thought experiment with various elements in order to describe movement of objects relative to other objects. Galileo illustrated this with a ship floating on a smooth sea with an observer, butterflies, a bottle dripping water and fish swimming in a bowl in a closed cabin. The observer will first observe these randomly moving objects while the ship is floating stationary and then while the ship is sailing. Here, there is no noticeable difference until the observer is allowed to look outside and see the direction of the ships motion. The movement of the fish, butterflies and even the bottle dripping water, all move relative to each other with no change even when the ship as a whole is moving or not, relative to land and sea, showing that all uniform motion is relative (Galileo, P. 187). This is the premise of Albert Einstein's Special Theory of Relativity and of General Relativity.

Another way to look at "relativity" is to look at how sound travels through air and the "Doppler" affect. For instance, let's say there is a train traveling at a constant rate with one observer on the train midpoint between the engine and the caboose. On the embankment next to the passing train is another observer, standing still, also midpoint between the engine and the caboose.

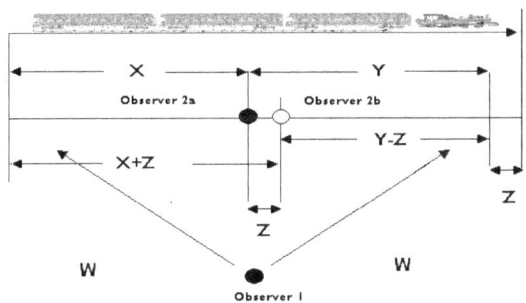

Einstein's Train - Sound

The engine sounds its whistle simultaneously with the caboose whistle (both engineer and conductor have their clocks perfectly synchronized). <u>Observer 1</u> on the embankment next to the passing

35

train (see Figure above) is equidistant from the engine and caboose as sound travels through the still air to the stationary observer (distance W); he therefore hears both whistles at the same time. Observer 2, on the train however, will have a different perspective. Observer 2 will move distance Z during the time it takes the sound to first reach him from the engine and caboose. The sound traveling to him from the engine will cover distance Y-Z while the sound traveling to him from the caboose will travel X+Z distance.

The speed of sound varies depending mainly on the, temperature and the *medium* through which it travels. At a standard atmosphere and at a temperature of 20^0 C sound travels at 344 meters per second in dry air or for a standard day, 59^0 Fahrenheit (15^0 Celsius), 761 miles per hour (Bord, 2000, p. 225). We shall therefore use 761 mph for the speed of sound in this mind experiment.

The train passing the embankment is traveling at 39 mph and is one mile long; therefore distances X and Y are a half-mile. 761 MPH = 12.7 miles per minute (MPM) or .211 miles per second (MPS). To convert miles per second to seconds per mile we inverse mps into seconds per mile (SPM) by 1/ .211 which is equal to 4.731 seconds per mile or 2.3655 seconds over a half-mile (Y).

The question now is how far did the Observer 2 travel during the time it took sound to cover distance Y and X? This would be distance Z. By simple calculation we discover that Observer 2 traveled Z distance in 2.365 seconds while traveling at 39 miles per hour (57.2 ft per second), therefore, 2.365 X 57.2 = Z = 135.278 ft. If half a mile (Y) = 2640 ft, then Y-Z = 2505 ft and X+Z= 2775.278 ft. If sound travels at .211 miles per second, it then travels at 1116 ft per second. Y-Z = 2.25 seconds and X+Z = 2.37 seconds a difference of .12 seconds.

Now, what if this was a high speed Bullet Train traveling at 139 mph (203.866 ft per second)? Therefore, Z distance for Observer 2 = 2.365 X 203.866 = Z = 482.14 ft. If half mile (Y) = 2640 ft, then Y-Z = 2157.86 ft = and X+Z = 3122.14 ft. If sound travels at .211 miles per second, it then travels at 1116 ft per second. Y-Z = 1.93 seconds and X + Z = 2.80 seconds, a difference of .87 seconds. In other words, Observer 2 at midpoint on the train will hear the engine whistle almost one full second before he hears the cabooses whistle.

The "Doppler Effect" is another phenomenon associated with sound and movement. We have all heard the train whistles pitch change as the train approaches us and then passes us – a higher to lower frequency.

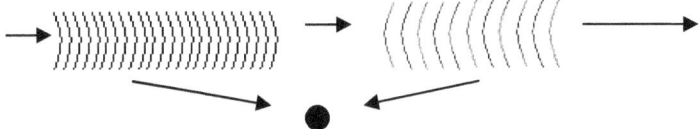

Doppler Effect on Observer 1

Although the Observer 1 on the embankment hears both whistles sound at the same time, he hears them at different pitches. As sound emits from the engines whistle, although its pitch is the same as the cabooses, its wavelength is lengthened or stretched as the train moves forward away from the observer, increasing the distance between wave peaks which in turn causes the pitch to be lower in tone. This is because sound transitions from a moving whistle into stationary air. Likewise the wavelength of the sound emitted from the caboose whistle moving towards Observer 1 on the embankment is compressed and shortened, increasing in pitch as the caboose moves forward into the still air.

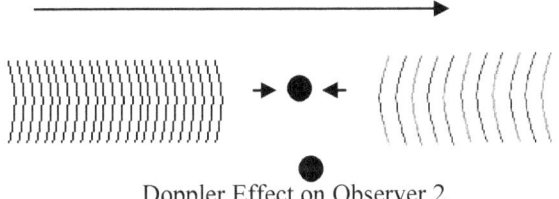

Doppler Effect on Observer 2

Observer 2 on the moving train however will hear the pitch from the engines whistle and the pitch of the caboose whistle at the same pitch, but not at the same time. Although the wavelength of the sound coming towards him has a longer wavelength, as he moves forward into the sound, the wave peaks reach his ears at a faster rate and he hears no difference. Sound also coming from the caboose, being compressed by the cabooses movement through still air will also be perceived as the same pitch as he moves away from the compressed sound waves, lengthening their perceived wavelength.

Einstein explains relativity similarly and quite eloquently with his "railway carriage" and "embankment" analogy in his book, Relativity, Special and General Theory (Einstein, Albert, Pg 10). In this comparison Einstein's train is also traveling at a constant rate along an embankment with observers on both the train and embankment. Lightning strikes simultaneously at the front and at the rear of the train while the observers on the embankment and on the train record their observations.

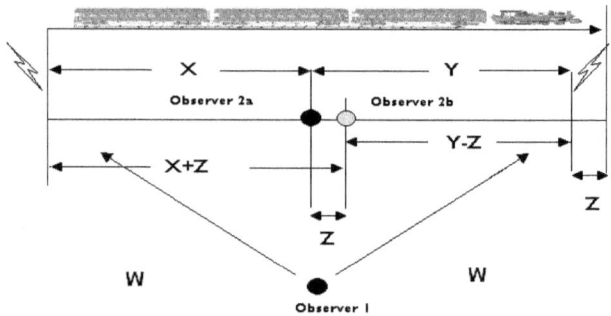

Einstein's Train - Light

The Observer 1 on the embankment records that the lighting struck simultaneously at the front and rear cars of the train, but Observer 2 riding in the center of the train records that the lighting at the front stuck first.

As the lightning struck, the train continued to move forward at a constant rate. By the time the light reached Observer 2 from the front of the train, the observer had moved distance Z forward to the Observer 2b position. The distance the light had to travel was then Y-Z. Likewise, Observer 2 moving Z distance away from the lightning striking at the rear car causes the light to travel distance X+Z to the Observer at the 2b position, therefore allowing the assumption that it struck later. This is exactly the same scenario and outcome as our thought experiment with the train whistle.

Rømer's Train

This same scenario (covered earlier with Einstein's Train) occurred with Dutch astronomer Ole Rømer in 1676 as he observed the movements of Io in orbit around Jupiter (Born, Max p. 91). Observer "A" (Romer) on the earth (the train) in the summer moved "Z" distance towards Jupiter while Io was in eclipse behind Jupiter; the light therefore had to travel Y-Z. Likewise in the winter (six months later) while the earth moved away from Jupiter, Observer B (Romer again) moved Z distance away from Jupiter during Io's eclipse, causing the light reflected off of Io to reach Observer B over a distance of X+Z. Remember that Cassini, whom Roemer studied under, observed that Io's eclipses got shorter as Earth approached Jupiter, and longer as Earth moved farther away.

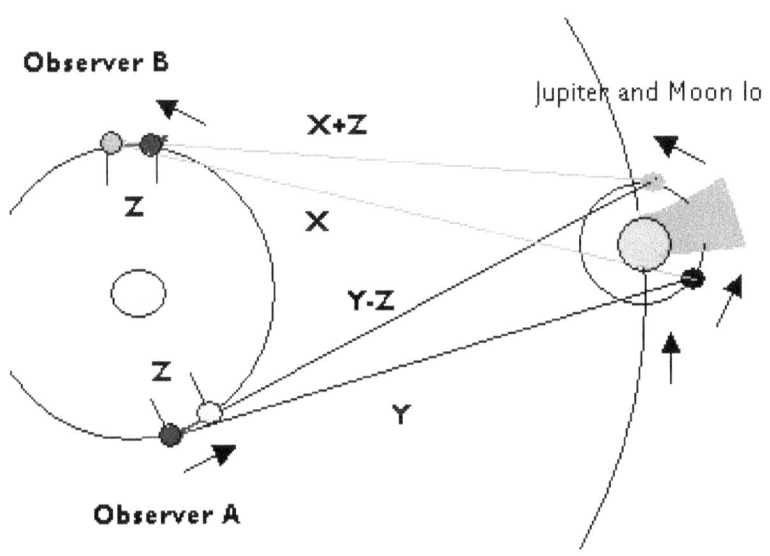

Jupiter and Io

Although this change shifted only seconds per day, to Romer this was a 17-minute change in Io's eclipse from summer to winter. This is due to the high rate of velocity of the earth towards Jupiter at one time of year (approximately 67,000 mph) while at the same rate of velocity away from Jupiter and moon Io months later.

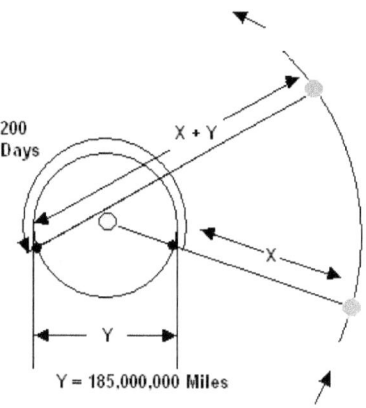

Rømer's Discovery

Prior to this point we have been focusing on the difference in velocity towards or away from an object (baseball, engine, caboose or moon). There is however an easier way to examine these changes. Keeping it simple, let's examine how Romer calculated the speed of light. At point X where earth is closest to Jupiter, Romer timed the orbit of Io around Jupiter. Two hundred days later when earth was at its furthest point away from Jupiter, Romer took a new reading which was 1,000 seconds longer. Scientists of that day had determined that the earth's orbital diameter was approximately 299,793,000 kilometers through triangulation and calculations performed on other planets. Romer determined that the additional time was due to the additional distance that the reflected light had to go – distance Y. If light took 1000 second longer to travel an additional 299,793,000 km, then through simple mathematics he calculated that the speed of light = 299,793,000/1000 seconds which is equal to 299,793 kilometers per second or 186,291 miles per second (Born, Max p.93).

Light, Gravity and Relativity

The theory of light prior to Einstein assumed that light was not made of matter and that gravity would have no effect on light. Einstein's Theory of Relativity and the field equations he had created generated a new assumption, that light was a particle and that being created of matter, gravity must attract light, curving its path as it passes close to a large stellar mass such as a star. A stars apparent position should shift by about 1.7 arc seconds according to Einstein's calculations due to the suns gravity (Bartusiak, Marcia, p. 47).

Einstein set out to observe and record how much a stars light was deflected by the sun due to the changes in space-time as light passed through its strong gravitational field. This was not an easy task as stars are not normally visible due to the suns brightness, so in 1919, astronomer **Arthur Eddington** took on this challenge and traveled to the Isle of Principle off the coast of Africa to observe this phenomenon during an eclipse. After studying the photographic plates, it became clear that the stars did shift their apparent position in the gravitational field of the sun. Gravitational forces did bend the rays of the starlight by about 1.6 arc seconds, proving that light was a particle that was made up of mass and that Einstein's theories were holding true (Isaacson, p. 257).

The figure below shows how the suns gravitational field pushes the apparent position of a star outward from its true position. We also see a similar effect due to gravitational lensing which also changes the apparent position of stars and galaxies due to massive objects throughout the universe.

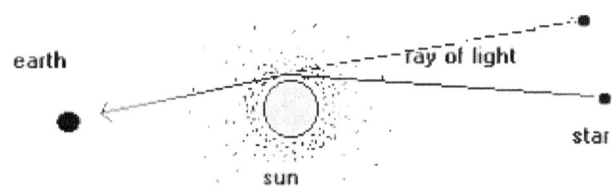

Apparent Position of a Star

Einstein developed his Principle of General Relativity, a unification of Newton's law of universal Gravity and his own Special Theory of Relativity. This theory would then go on to predict gravitational lensing, gravitational redshift, the precession of Mercury and many more scientific phenomenon discovered later in the 20th century. This principle was generalized in a mind experiment where passengers on a train moving along with uniform motion feel a sudden change in motion with the abrupt application of the brakes (Einstein, Albert, p. 73). Einstein would go on to develop the **Equivalence Principle** from this showing that inertial mass that is produced by acceleration/ deceleration and gravitational mass in a gravitational field is essentially the same thing.

Gravitational Lensing

While writing for a science yearbook in 1907, Einstein noted a thought experiment of an observer in a closed windowless chamber that is accelerated upward. He compared the acceleration of this observer to another observer in a gravitational field showing that both felt pressed to the floor and that neither could distinguish the difference, thus formulating his "equivalence principle." This principle asserts that inertial mass and gravitational mass are equivalent, that both resistance to acceleration and gravitational effects are manifestations of inertio-gravitational fields.

In this same sort of scenario, an observer is in a chamber accelerating upward, again utilizing the "equivalence principle." A pinhole in one wall allows light from outside the chamber to shine on the opposite wall of the chamber. By the time light hits the opposite wall the chamber has moved causing the light to hit the wall a little closer to the floor, in effect **bending the light**. Since inertial effects are equivalent to gravitational effects, gravity too should cause the light to bend (Isaacson, Walter p. 190). This is strangely similar to James Bradley's Star Aberration reviewed earlier with the "Tall Box Analogy".

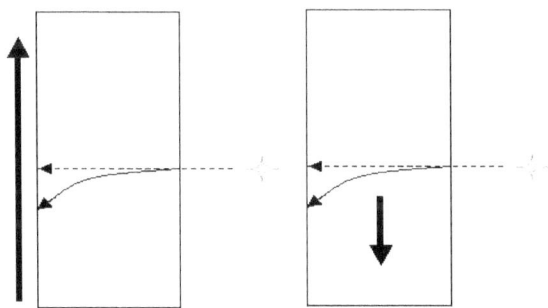

Equivalence Principle

Quasars, short for quasi-stellar radio source, were first discovered in the 1950's well after Einstein's theories were developed. Quasars are very powerful electromagnetic wave generators at the center of distant galaxies that are much brighter than normal galaxies. Quasars are the strongest sources of electromagnetic radiation in the universe, possibly generating energy from a super massive black hole at the center of its galaxy. Because of their great distances and brightness, quasars are ideal for studying the effects of gravitational fields on light. Einstein theorized that light would follow a curved path around massive objects due to the immense gravitational field. Today many stellar objects such as Einstein's Cross or Einstein's Rings occur due to the deformation of the light from a source such as a quasar through gravitational or cosmic lensing.

Gravitational lensing occurs when the light from a quasar or other very distant and bright source curves around an enormous object, such as a galaxy, that is between the quasar and the earth. In 1979 a "Double Quasar" named Q0957 was discovered with the first confirmed case of gravitational lensing. This double quasar was caused by the refraction like effects of intense gravitational fields bending the light of a single quasar into two observed quasars (Bord, p. 498). More than 30 such gravitational lenses are known today including the **Einstein Cross** (G2237 + 0305 - on the cover of this book) where one quasar is refracted into four images of itself through gravitational lensing from a huge galaxy (Dekel, p. 383).

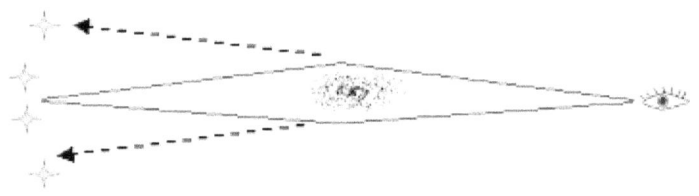

Gravitational Lensing

Gravitational Red Shift

The movement of the earth relative to a star or galaxy or the reverse, galaxy or star towards the earth produces a strange phenomenon known as **red shift**. Armand Fizeau, (inventor of the spinning toothed wheel experiment) described this spectacle in 1848 after noticing a shift towards red in the spectral lines of the light being viewed from stars moving away from the earth. Willem de Sitter had also noted this phenomenon in 1913 with the twin star Doppler Effect covered earlier. **Red shift** can be defined as a **stretching** in the wavelength of electromagnetic energy detected by an observer or detector compared with the original wavelength emitted by the star **moving away** from the observer. This expanded wavelength is also directly proportional to the drop in the frequency of the radiated light. The opposite is also true with the **compression** in the wavelength of light coming from another object **moving towards** the observer and/ or detector, commonly called **blue shift**.

Edwin Hubble's (1898-1953) earth-shattering discovery in 1929 suggests that **galactic red shift** in light from stars and galaxies increase in proportion to their distance from earth, this is known as Hubble's Law (Ditchburn, R.W., Pg 340). Studies show that 38 of 40 galaxies observed are rapidly moving away from each other. This omni-directional galactic expansion shows that galaxies furthest away are moving the fastest away from earth– some at 57,000 kilometers per second or 13 million miles per hour – 2% the speed of light (Bord, p. 220). This process implies that some sort of repulsive force exists between the galaxies, similar to an expanding gas or substance that is pushing galaxies outward and away from each other – somewhat like bread rising. Gravity in effect should be pulling galaxies together in what ultimately would be the "Big Crunch". This possibly is why galaxies further away from each other are moving faster away from each other – gravitational forces are much weaker. Further research also shows that this expansion is accelerating, additionally implying that expansion agents also known as **Dark Energy** and **Dark Matter** (two different substances) exists in the universe. This substance is expanding between the galaxies and stars, forcing them outward, away from each other against their own gravitational attractions (Bord, p. 493). Empty space may not be so empty as it may seem.

Gravity was now known not only to be able to bend light but even slow it down. A similar phenomenon known as **gravitational red shift** is noted in light emitted from very large stars, which is not related to movement but due to the immense gravitational fields of the star itself. The gravitational force of the star pulls at the light emitted from the star, slowing it down as if it had to climb up steep hill. This slowing effect on the light coming from the star causes the wavelengths of that light to become stretched, shifting the light toward the red end of the spectrum, called gravitational red shift. This red shift is only possible if the gravitational pull on the light emitted from the star is greater than the gravitational pull of the earth where the observer and detector are as the earth gravitational force has the opposite effect, speeding up the light (Ditchburn, R.W., p. 315).

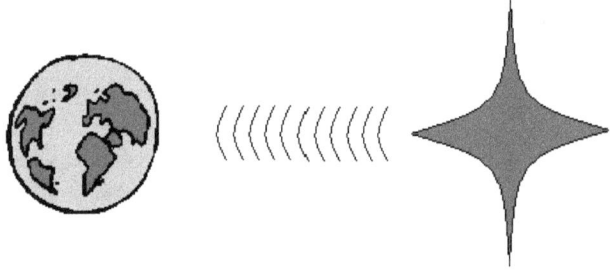

Gravitational Red Shift

Light reflected off the moon however, has the opposite effect, shifting towards the blue end, known as gravitational blue shift. This is only possible if the gravitational pull on the light reflected off the moon or any other object is less than the gravitational pull of the earth where the observer and detector are (Isaacson, Walter, p. 148).

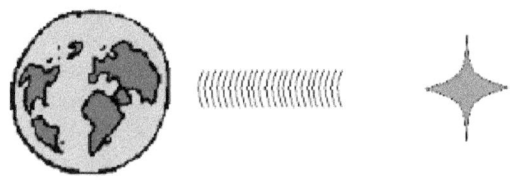

Gravitational Blue Shift

Highly accurate clocks onboard 24 **GPS satellites** circling the earth at 20,000 km (12,427 miles) altitude also are affected by red shift causing another phenomenon believed to be time dilation, a warping of space and time as predicted by special relativity (Bord, p. 13). Gravitational red shift has a positive effect with GPS due to the weaker gravitational field of the satellite relative to the earth's surface. Time dilation however, has a negative effect due to the velocity of the orbiting satellite of 4 kilometer per second or 8,948 miles an hour relative to a stationary point on the earth, both tending to cancel each other out.

These GPS clocks must be synchronized to be within 50 billionths of a second in order to be capable of measuring positions on the ground within 4.5 meters (15 yards). These clocks must be corrected for "relativistic anomalies" due to gravitational red shift, otherwise they would run faster by more than 40,000 billionths of a second each day, changing the calculations of positions on the ground. Without these corrections, the clocks onboard the 24 satellites would be out to sync within a minute and a half – a matter of national security (Bartusiak, Marcia, p. 55). Einstein predicted that clocks would run slower in more intense gravitational fields (Isaacson, Walter, p. 148). It is the earth's gravitational field that affects these clocks; therefore clocks at different altitudes should run at different rates having their frequencies shifted (gravitational redshift) due to gravitational differences. Two identical clocks at different altitudes will therefore run at different rates, with large enough difference to cause a variation in distance on the ground by as much as 13 km in a day.

Precession of Mercury

Another orbital phenomenon is the orbital precession of the planet Mercury. Mercury's orbit is an extremely elongated elliptical orbit approximately 36 million miles from the Sun that precesses ever so slightly every orbit. Newtonian formulas explained this phenomenon by taking into effect the gravitational forces of all the other planets on Mercury including the effect from the Sun and its own rotation. Using these calculations, scientists calculated that Mercury's perihelion (the closet point of its orbit to the Sun) should shift or precess about 531 arc seconds per century. In reality, Mercury precesses about 574 arc seconds per century, 43 arc seconds more than the Newtonian predictions (Hobson, p. 233).

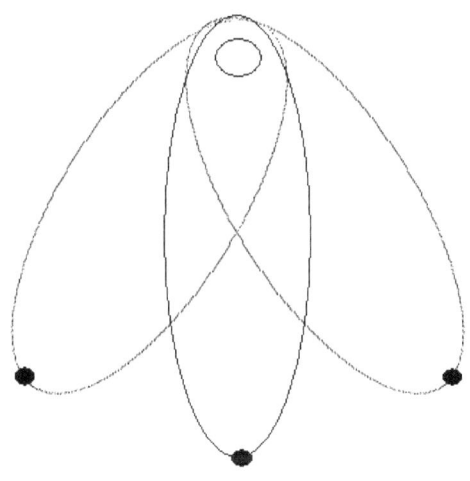

Mercury's Perihelion Shift

Newtonian formulas could not account for this change in motion. Something invisible was shifting the orbit ever so slightly, orbit by orbit. It has been suggested by some scientists that another planet, some call it **Vulcan**, may be in an orbit slightly closer to the Sun than Mercury, producing an additional gravitational force off center with the suns gravitational grip that could account for this discrepancy. Some other scientists have also theorized that there must be a certain amount of dust between the Sun and Mercury, which puts a drag on its motion and changes its orbit. No evidence however of dust or another planet has ever been found in this region. Einstein in 1915, calculated that the orbit of Mercury should precess by about 43 seconds of arc per century using the General Theory of Relativity, exactly what it does (Bartusiak, Marcia, p. 42).

Mercury's orbit ranges from 46 to 70 million km from the sun due to its elliptical orbit. Mercury is the fastest-moving planet in our solar system with a mean orbital velocity is 47.87 km per second with a minimum velocity of 38.86 km per second and a maximum velocity of 58.98 km per second, twice that of the earth. As Mercury changes velocity and enters into a stronger gravitational field during its Perihelion or closest encounter with the sun, its mass increases and its local time changes ever so slightly. These slight changes in mass and time due to the General Theory of Relativity help account for the missing 43 seconds of arc per century.

Tracking Venus and Mercury

If gravity of a large body in space such as a star could produce time dilation or a warping of space-time bending light rays, theoretically the sun could also produce this phenomenon. Working with astronomer Arthur Eddington in 1919 while viewing a solar eclipse, Einstein showed that the gravitational field of the sun actually did bend rays of light from a star. The position of the star was known but as its light rays passed close to its atmosphere, a form of Gravitational Redshift occurred, shifting its apparent position (Isaacson, p. 257). The question arose that if the suns strong gravitational field affected light, could this same phenomenon be also evident with other electromagnetic energies?

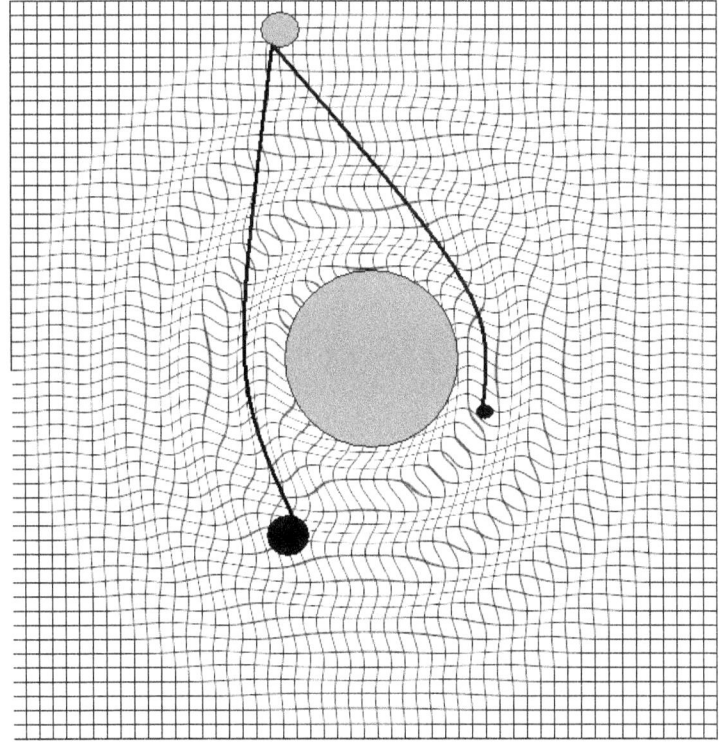

Radar Tracking of Mercury and Venus

In experiments carried out over several years beginning in 1964, Harvard Smithsonian astronomer **Irwin Shapiro** discovered a new variation of Gravitational Red Shift. Using **MIT**'s newly built Haystack Observatory located in Westford, MA, Shapiro started tracking and recording Mercury and Venus's positions when earth and each of these planets were on opposite sides of the sun. By bouncing high-frequency radar signals off the planets, Shapiro found that as the planets started to pass behind the sun, the return time of the radar signals increased by about one five-thousandth of a second. This lengthened the 23-minute round-trip time to Mercury and the 30 minute round trip time to Venus of these radar signals, simulating an increase in the distance of the planets orbits as if the radar beam had bent or "dipped" changing course, increasing the distance it traveled to and from the planets (Bartusiak, Marcia, p. 51).

Cosmic Background Radiation

In 1964 while working on ultra-sensitive cryogenic microwave receivers, Bell Laboratory researchers Arno Penzias and Robert Wilson came across an unexplained microwave radio noise (Bartusiak, Marcia, p. 63). This pesky radio noise was isotopic, meaning it came from all directions, far weaker than the radiation given off by the Milky Way. They first assumed that their instrument was picking up interference by local radio and TV station noise from New York City but noticed that it came from all directions averaging in the microwave range of approximately one millimeter. The next assumption was that a pigeon dropping in the radio antenna's large horn was causing the microwave radio noise. After cleaning up the pigeon mess, the noise still frustratingly remained (Bord, 2000, p. 304). Further research showed that this radiation was from so far away that it must have been generated over 15 billion years ago, a half a million years after the "**Big Bang**", before the formation of the planets when the universe was filled with a red-hot mist of plasma.

Optical radiation began to be generated as the first atoms began to form; the "primordial fog" was then lifted and the universe became relatively transparent, beaming with light (Bartusiak, Marcia, p. 226). Prior to the formation of atoms, the universe was probably amassed with only the building blocks of atoms: protons, electrons and neutrinos. Research on this newly found phenomenon shows that galaxies are all rushing away from each other at very high velocities and seems to show that the universe is constantly expanding (Ditchburn, R.W., p. 340). It is believed that as the universe expanded, the hot plasma evenly distributed through the universe cooled. The light that was being generated at that time, mainly in the optical and infrared spectrum, became stretched, increasing its wavelength and lowering its frequency down from the optical range, down into the microwave range. While viewing the universe today with a traditional optical telescope, the background of space is now totally black. With a radio telescope however, there still remains a weak glow in all directions in the microwave radio frequency range of 160.2 GHz; this microwave radiation now being detected is scientifically known as **Cosmic Microwave Background Radiation or CMBR.**

Could this energy at the edge of our known universe be the remnants of the plasma that was generated during the Big Bang? Could this background noise actually be the energy of the material that all particles condensed from? If so, this relic that we see is the way it was 15 billion ago, which has by now has condensed into stars and galaxies, a universe we can never really see.

Dark Energy and Dark Matter

As previously mentioned, **Dark Energy**, the hypothetical expansion agent, possibly causes the rapid dispersion of our universe. As particles of matter gather great distance between themselves, gravitational attraction between them begins to wane, allowing this expansion agent to more rapidly separate particles from each other, accelerating them exponentially (Hawking, p.48). The universe is believed to be composed of 73% of this mysterious **Dark Energy**, 23% of an unknown type of **Dark Matter**, and 4% of ordinary matter. Through understanding "Hubble's Constant" and calculating the known visible mass of galaxies, the universe should be expanding at a specific rate, but it is not. Through measurements of redshift in the light emitted from these galaxies, the universe is expanding much faster than first believed. The only way to explain this is to calculate in the missing mass (Glendenning, p. 23).

The measurement from the Wilkinson Microwave Anisotropy Probe (WMAP), which was launched in 2001, has played a primary in developing today's Standard Model of Cosmology. WMAP, through the study of Cosmic Background Radiation has also determined that the cosmos is relatively flat and 13.7 billion years old (Glendenning, p. 64).

This Background Radiation was determined to be homogenous or proportionally the same throughout the universe as a "smooth sea" of energy even though the stars and galaxies are clumped together in large masses with "nothing" in between them (Bartusiak, Marcia, p. 63). This energy seems to act as a hot gas, expanding outward, pushing everything around it away. As larger distances separate galaxies, the gravitational pull that attracts them diminishes, allowing this expanding "gas" more leverage, accelerating these masses faster and further apart. This was Einstein's **Cosmological Constant**, which represented a repulsive force in the universe, counteracting gravitational forces that attract bodies through out the universe (Isaacson, Walter, p. 353). This constant was developed in order to maintain a balance in the universe, preventing it from collapse – a static view. After Hubble's discovery that indicated an expanding universe, this equation was modified, allowing for Hubble's Law (Ditchburn, R.W., p. 340).

In the 1930's, Cal Tech professor **Fritz Zwicky** was first to conceive of **Dark Matter** in an effort to account for the high velocity of stars in the outer fringes of our own galaxy. Theoretically, these stars would spin off into space due to centrifugal force, but an unseen mass needed to be added, as gravity due to the visible mass of luminous stars was not enough. Dark Matter can also be used to account for gravitational lensing that curves light rays passing through the immense gravitational fields of galaxies. There has to be much more mass exerting gravitational forces than was seen visibly. The majority of Dark Matter is believed to be non-baryon, meaning that it does not have the atomic structure of visible mass such as the hydrogen atom and may not be able to interact electromagnetically with discernible mass other than gravitational forces (Glendenning, p. 49). Dark Matter would evidently not be associated with large **black holes** in much of the universe, as gravitational lensing would indicate their presence, however, the existence of miniature black holes permeating the galaxies proportionately dispersed, is still a possibility.

The Muon Life Cycle

Muons are highly unstable subatomic particles with a life span of 2.2 microseconds (.0000022 of a second) and are only produced under intense energies. Muons like electrons are classified as leptons and have a mass, which is about 200 times that of the electron (Bord, 2000, p. 451). Outside of nuclear events, muons are only naturally created by cosmic radiation composed of protons traveling at near light speed from deep space impacting air molecules at the edge of space. Given their short life span, muons should never reach the ground, decaying into an electron, an electron-antineutrino, and a muon neutrino high up in the atmosphere (Bartusiak, Marcia, p. 34). Mathematical calculations predict that less than one muon out of a million should ever hit the earth's surface. Experiments however show that about 50,000 out of a million muons created in this exchange hit the ground and that 10,000 muons hit every square meter of the earth every minute of every day – how could this be?

Only a relativistic resolution can be found. It is believed that muons traveling at .98 times the speed of light undergo time dilation where local time on the muon is much slower or length contraction occurs which allows for them to exist long enough, 2.3 times its life span, to hit the ground (Bord, 2000, p. 452).

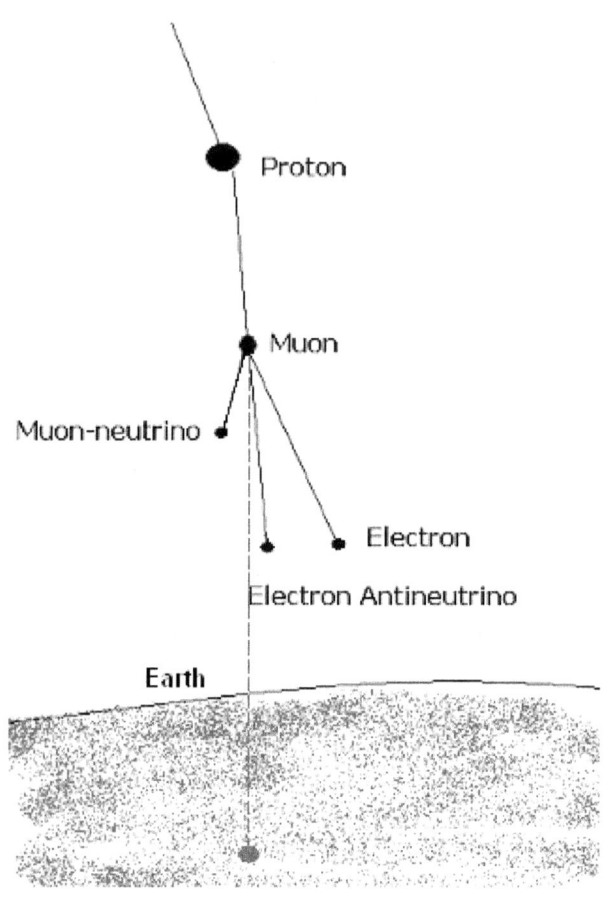

The Muon Life Cycle

Relativity

The "Lorentz Transformation" Hypothesis

The failure of Michelson and Morley to detect the aether was puzzling. Soon after the Interferometer experiment, Irishman **George F. Fitzgerald** (1851-1901) professor of experimental philosophy suggested that the negative results of the experiment might have been due to the contraction of the interferometer in the direction of movement. Hendrick Lorentz built on this idea developing the contraction hypothesis based on his Electron Theory stating that any material body is contracted in the direction of its motion. This contraction would then change by a factor relative to the speed of light (Einstein, Albert, p. 10).

Lorentz's contraction theory would soon serve as the foundation for the mathematics of Einstein's theory of relativity. Einstein developed on this hypothesis making some substantial changes. Einstein modified the Contraction Theory showing that **these changes due to motion do not actually warp the body in motion**, as Fitzgerald and Lorentz had originally suggested, but instead warped or **made changes in space and time around that moving body** (Ditchburn, R.W. p. 318).

Lorentz's transformations, which he introduced in 1904, formed the basis of Einstein's Special Theory of Relativity. They describe the increase of mass, the shortening of length and the time dilation of a body moving at speeds close to the velocity of light. This may all seem out of the norm as no one or nothing man has yet made even travels close to these speeds. However, this does affect our Earth and us on the Earth. The Earth moving in its orbit around the Sun moves at about 30,000 meters/sec (30 km/sec or 18.5 mile/sec). The contraction of space caused by the high velocity of Earth through space would contract the diameter of Earth by about 6 cm (2.5 inches). This small change may account for Michelson and Morley's negative result by making the source of light and the mirror draw closer together when the system is moving lengthwise nullifying any change due to the earth's rotation and velocity through space.

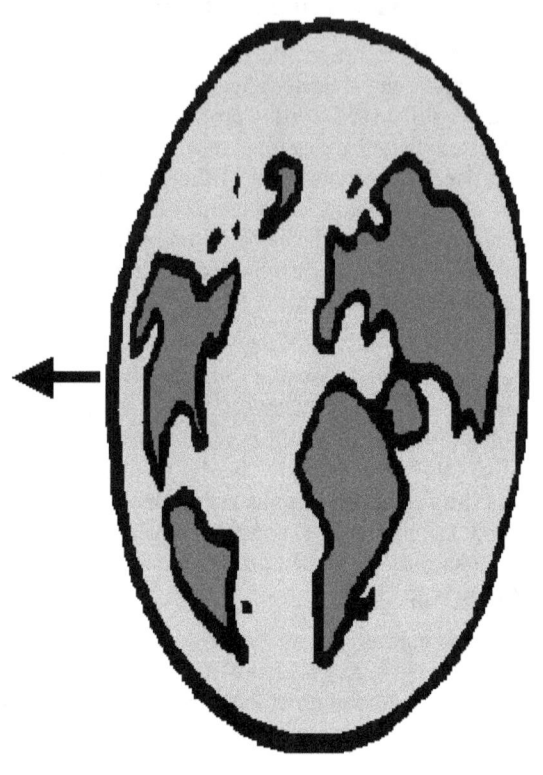

Length Contraction of the Earth

Lorentz Transformation of Time and Space

In three-dimensional space, the Moving System moves with velocity v in the x direction with respect to the Fixed System reference; the moving X' is contracted.

"z" represents height, "y" represents width and "x" represents length of an object. Here, measurements are compared to a fixed system relative to a moving system. Note – Length of x is not = x'.

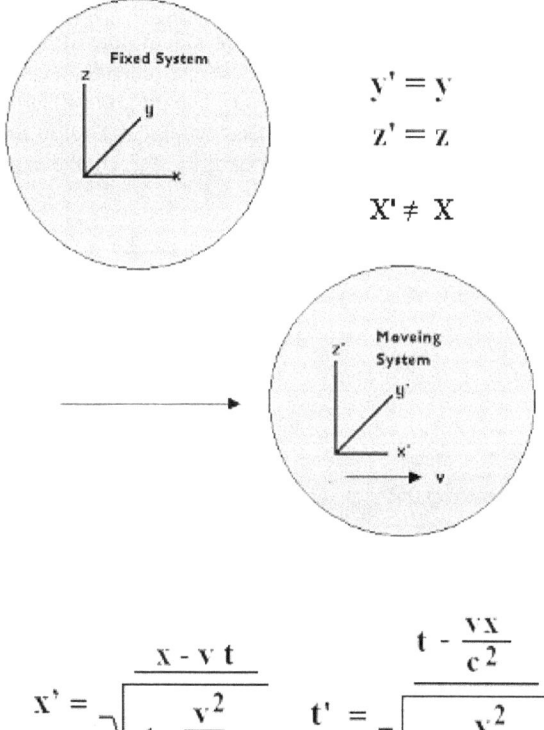

$$y' = y$$

$$z' = z$$

$$X' \neq X$$

$$x' = \frac{x - v\,t}{\sqrt{1 - \frac{v^2}{c^2}}} \qquad t' = \frac{t - \frac{v\,x}{c^2}}{\sqrt{1 - \frac{v^2}{c^2}}}$$

Lorentz Transformations

Lorentz used an arcsine or inverse sine formula to express these transformations.

$$= \sqrt{1 - a^2/b^2}$$

This formula divides the square of two values (such as two velocities) in order to show the ratio between these two values in a range from 0 to 1 (**A**). This ratio is then subtracted from one to return an inverse nonlinear value that cannot be greater than 1 (**B**). This value is then factored with other values that determine the change in length, time or mass variable that changes according to velocity (**C**). The number one or maximum values is then substituted with the value of the speed of light, 186,000 miles per second or 300,000 kilometers per second; the percentage can also be used such as .8 to express 80% light speed.

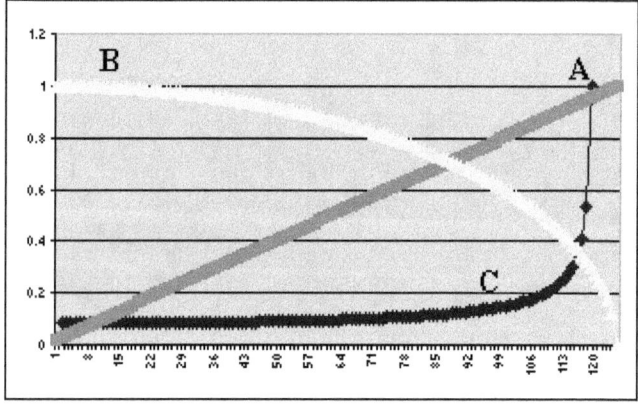

Formula Variations

Examples:

Fixed System

v = velocity = 0
l = length = 100 ft
t = time = 10 seconds
c= speed of light
m = mass = 10 Newton's

Moving System 1

If velocity v = .5 c then

Length of contraction = .866 x 100 = 86 ft

Time dilation =   1.15 x t = 11.5 seconds

Relative Mass = 1.15 m = 11.5 Newton's

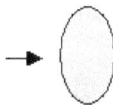

Moving System 2
If velocity v = .999 c then

Length of contraction = .044 x 1 00= 4.5ft

Time dilation =   22.366 x t = 3.7 minutes

Relative Mass = 22.366 x m =223.7 Newton's

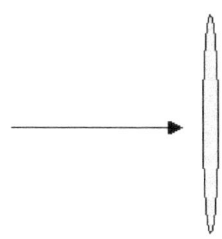

$$l' = l \sqrt{1 - \frac{v^2}{c^2}}$$

Contraction of Space

The "Lorentz Contraction" Theory (Einstein, Albert, Pg 53)

Length Compaired with the Velocity of an Object

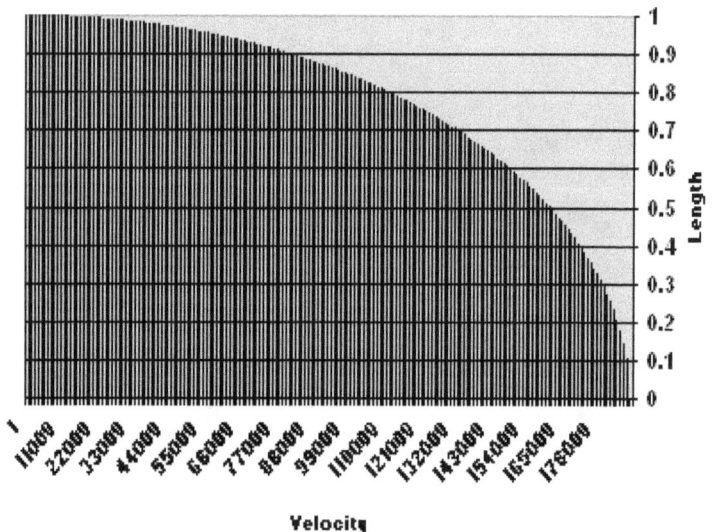

Length Contraction

$$t' = \frac{t}{\sqrt{1 - \frac{v^2}{c^2}}}$$

Dilation of Time

The Lorentz Time Dilation Theory

Time Compared with the Velocity of an Object

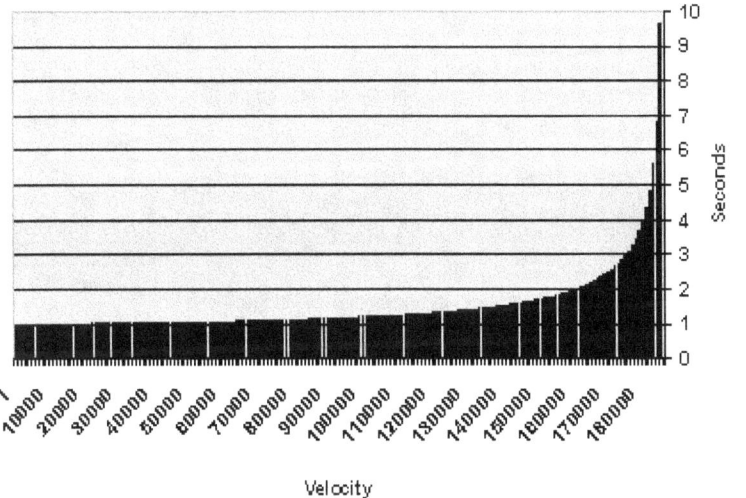

Time Dilation

$$m = \frac{m_0}{\sqrt{1 - \dfrac{v^2}{c^2}}}$$

Relativistic Mass Formula

Mass Change in Velocity

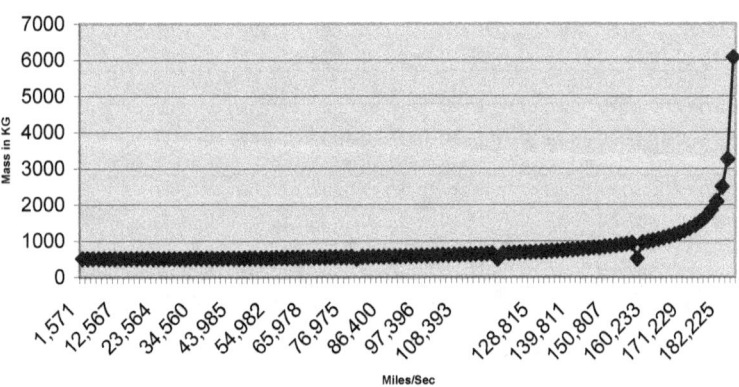

Relativistic Mass

Einstein's Remedy

As we saw earlier, Einstein's remedy for this dilemma in the early twentieth century was to alter time and space, building on **Hedrick Lorentz's** Contraction Theory. As discussed previously, if a baseball is flying at 100 mph and is measured by a radar gun that is moving in a vehicle towards it at 50 mph, the reading on the radar gun would read 150 mph. The only way to keep measuring the velocity of the baseball at 100 mph instead of 150 mph is to alter "**time and space**." We would therefore either have to change the length of the measuring rod, as Einstein would say it, or change the rate of the clock or both. In this case we would have to change the way the radar gun calculates the rate of change in the baseballs position.

This of course does not affect our baseball game; baseball speeds just aren't fast enough. This concept is however very important to understand as we venture further into the future and develop new technologies. If we ever do develop these hyper speeds, what would happen to us at these velocities? According to Einstein for example, if the measuring rod on the moving object was a yard, it would then have to be perceived to be shorter by a stationary observer (Contraction of Space) and one second to the stationary observer would be less than one second on a clock on the moving object (Dilation of Time). Of course inside that moving system the yard would still be perceived as one yard and one second perceived as one second; no change at all would be perceived in space and time by an observer moving with that system (local space and time). In that way, to a moving observer, no matter what their velocity towards or away from the light source, their readings while detecting the speed of light will always read 300,000 km/sec or 186,000 miles/sec.

In theory, velocity changes space and time. The faster the velocity of an observer, the slower he moves forward into time compared to a stationary observer. Below is an illustration of Einstein's Co-ordinate Systems A and B. System "A" represents a stationary system and System "B" represents the moving system. The embankment - (System A) in Einstein's Train analogy symbolizes the stationary system while the train represents the moving system (System B) (Isaacson, Walter, p. 123). The height or vertical plane of three-dimensional space in both systems is represented by y where y' = y. The horizontal aspect of width in both systems is represented by z where z' = z. The horizontal plane that corresponds to the fore and aft movement in three-dimensional space in both systems is represented by x where x' = x – velocity multiplied by time or x-vt. The length of an object is shorter or compressed in the direction of movement by a magnitude the faster an object moves as seem in length x' relative to x.

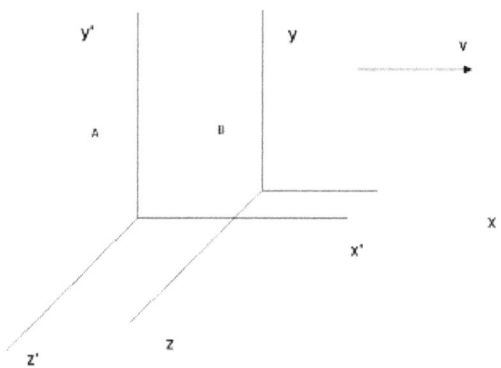

Co-ordinate Systems – Contraction of Space

Illustrated below is the variation in time (Local Time) between three moving systems traveling at three different velocities. System "A" is traveling at 1000 mps (miles per second relative to a stationary observer) where 1 second is equal to 1.000014 seconds. System "B" is traveling at 170,000 mps where 1 second is equal to 2.464495 seconds. System "C" is traveling at 180,000 mps where 1 second is equal to 3. 969143 seconds. Time moves slower in a moving system than a stationary system, therefore theoretically, the faster an object moves the slower a clock ticks. At a very high velocity, say on a spaceship, this time dilation would make it possible for passengers to travel further into the future while aging very slowly compared to people on a stationary object (Born, Max, p. 317-320). In other words, everyone on earth has aged much more than those in the spaceship

Seconds	Velocity
1.000014	1,000 mps
2.464495	170,000 mps
3.969143	180,000 mps

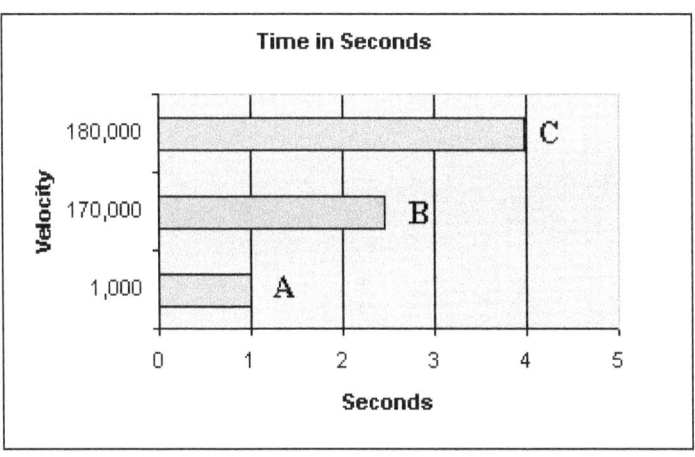

Co-ordinate Systems – Dilation of Time

Blast Off

To test this theory, let us perform what Einstein called, a thought experiment. When? It is the year 2050 and technology has brought us to a turning point. Using a now safe Cold Fusion nuclear power plant and the latest version of the Ionic Thrust Engine (ITE), we are now ready for a venture never before possible. Assembled in orbit around the earth, a five-stage space vehicle is readied for launch with five crewmembers on board. Its fuel is water, accelerated in vapor form to approximately the speed of light by ionizing droplets of the water and accelerating them though a magnetic field. Each stage will "burn" for approximately twenty-four days straight, maintaining an acceleration of 98 ft/sec^2 and a constant force on the crew of 3g's (three times Earths gravitational force) for the entire burn. If you would experience this in a car, this would be an acceleration of approximately 0-66 miles per/hour in one second maintained for twenty-four days straight, a little rough on the human body.

The crewmembers are strapped in and hooked up to devices that provide nutrients and also supply and remove fluids and toxins from their bodies over the next twenty-four days. After blastoff, the crew monitors the spacecraft for a few hours, and then is put into a deep sleep. After these twenty-four days are up, the crew are gently awakened and allowed to adjust to the environment. Then on the twenty-fifth day, engines are shut down and the acceleration stops – the crew are weightless. Crewmembers unbuckle themselves and enjoy this release of pressures on their bodies and are give a day to recuperate.

At the end of their one-day break, one crewmember enters the observation cabin on the first stage booster. The booster is then separated from the space ship; the crew is readied, and then blasts off again leaving Stage One and one observer behind. Like the first stage, this ship also accelerates to 3 g's for twenty-four days straight and the crewmembers again sleep away the time until awakened for the next brief rest. Like previously done, the second stage and one observer are left behind and Stage Three blasts off, accelerating on at 3 g's for twenty-four more days off into space.

This process goes on five times till the fifth stage and its lone observer shut down the engines and cruise weightless in space in a different world. Where are they and how far and fast have they traveled? Let's find out.

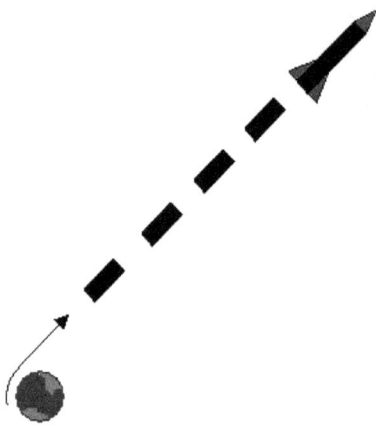

Five Stage Rocket to Light Speed

First burn - How far did the crew travel and how fast are they traveling having maintained 96 ft/sec/sec or 3 g's for twenty-four days straight? Remember, this was 0-66 miles per hour in one second. After one minute or 60 second later, the crew would be accelerated to 3,927 miles per hour and after one hour, 235,636 miles per hour. When twenty-four hours or one day has elapsed, the space ship will be traveling at 5,655,272 miles per hour or 1,570 miles per second. When the first twenty-four days have past and the crew is on their first well needed break, they will be traveling at 37,701 miles per second, one fifth the speed of light.

Second Burn - This next stage blasts off and accelerates for twenty-four more days up to 75,403 miles per second. Third through Fifth Burn - this process continues until the fifth stage completes its burn traveling at a calculated velocity of 188,509 miles per second, faster than the speed of light in reference to the Earth. A total of 121 days have now passed and the spacecraft has traveled more than 60 billion miles.

Now that we have gone this far, it's time to return back to earth. Each stage of this ship adrift in space retained half of its fuel (water) for the return home. The fifth stage and its resident take one day to recuperate and then turn the ship around, firing up the engines for the long return trip back home. Blasting off back towards earth, the fifth stage returns back toward the direction it came from, with a rendezvous of course in store with Stage Four floating aimlessly in the deepest parts of the universe.

Like all of the other separation launches of each stage, Stage Five also accelerates up to and maintains an acceleration of 98 ft/sec^2 or 3 g's. This acceleration however is back in the direction of which Stage Five came from and is in fact a deceleration, facing in the other direction. The effect on the ship and its passenger is however exactly the same, being pinned up against the seat for 24 days straight. What we are doing is actually slowing down Stage Five to Stage Fours velocity and catching back up to it to rejoin the capsule and crew. This will go on for four more link ups as the space ship returns back to earth, taking another 121 days to complete. Each launch decelerates the ship down to the velocity of the prior stage until the ship returns back to earths orbit whole again.

If this is hard to understand, let's put it this way. When Stage Five blasts off and leaves Stage Four behind, it will be traveling after 24 days at 37,701 miles per second away from Stage Four. In space however, this is all relative. Who is to say that Stage Five is moving away from Stage Four? Stage Four could also be viewed as moving away from Stage Five at the same velocity with Stage Five being stationary in space (after it stops accelerating of course). If this were the case, Stage Five would have to use its rocket engines to catch up the Stage Four. This is in fact what we are doing on the return back to earth. With no real reference of what is stationary space, anything not accelerating or decelerating can be considered as stationary. Each stage viewed from another will be traveling at a different velocity – see Relative Velocity chart.

Now, according to calculations, the observer on Stage Five before his return on day 121 was theoretically traveling at 188,509 miles per second relative to the earth. Stage Five would be therefore traveling 2,218 miles per second or almost 8 million miles per hour faster than the speed of light. Can this be?

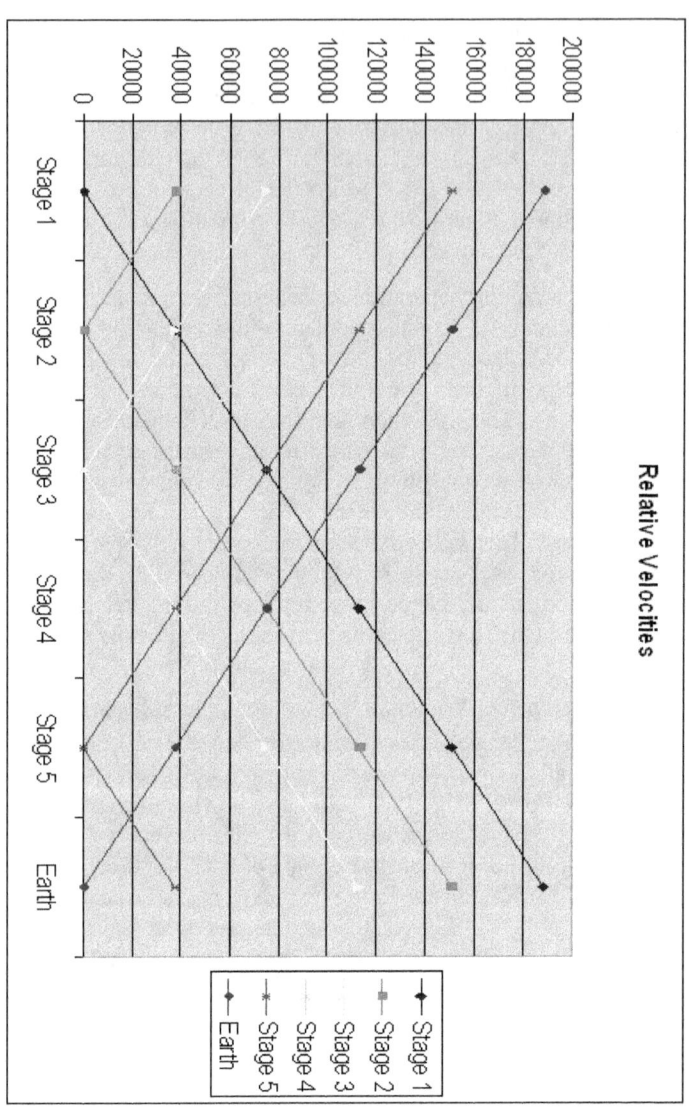

Relative Velocity Graph

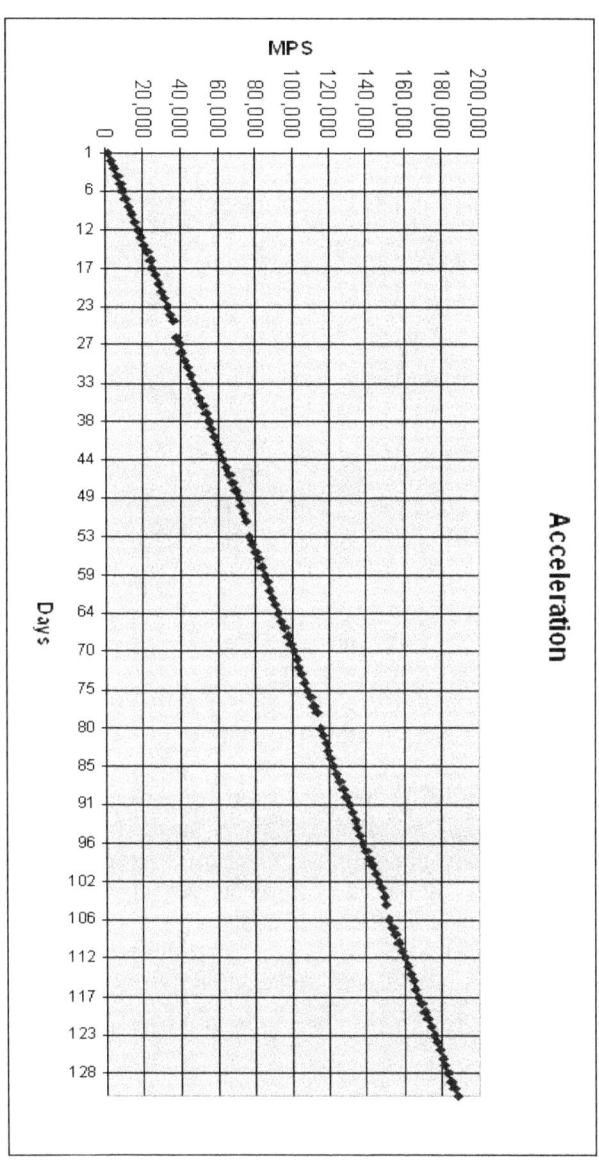

Acceleration Graph

Theoretically nothing can go faster than light. However, if the first stage of a two stage rocket reached velocities close to the speed of light say .95 times the speed of light, and launched the second stage up to the same velocity away from Stage One, you would have this scenario.

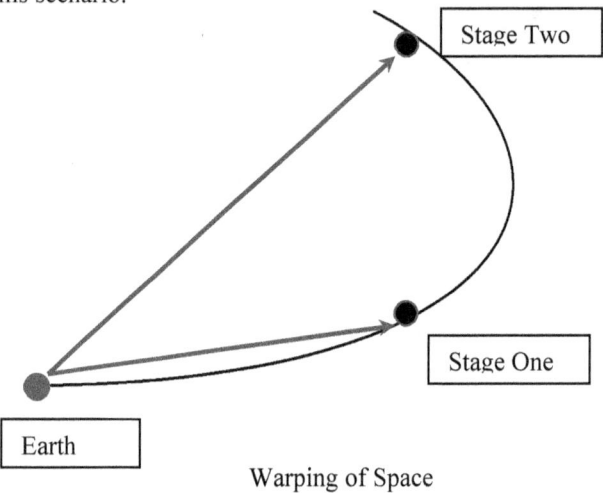

Warping of Space

Einstein's Theory of Relativity would "warp" or bend space in order to prevent Stage Two from going "Warp Speed" or the speed of light. From the Earth's perspective, if Stage One was at .95 the light of speed, Stage Two could not go very much faster, say .98 the speed of light and would have to travel in a curve. The Lorentz Contraction Theory introduced by Einstein does very much the same thing by reducing the rate of change in distance exponentially. This rate of change occurs the faster an object moves away from or towards an observer until the velocity does not change at all, preventing an object from reaching "Warp Speed."

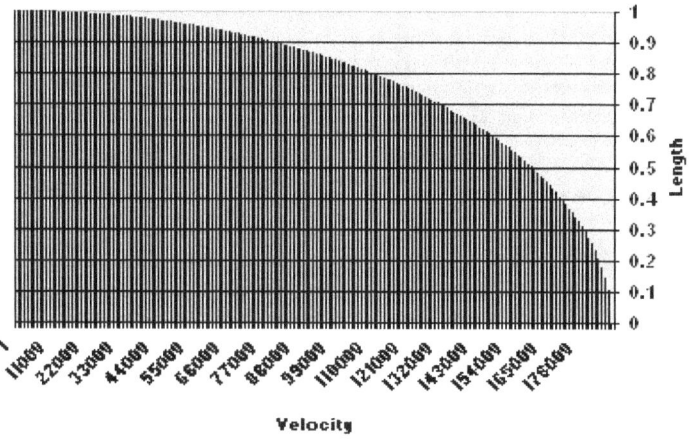

Length Compaired with the Velocity of an Object

Velocity

Contraction of Space

From the earth's perspective (see chart on following page), with each stage floating in space moving at different velocities relative to the earth, lengths will contract more and more. For example, with a ten-stage spaceship, stage one will be viewed as having very little change. Stage five however will have a contraction of approximately 80% earth's measurements and stage 10 with length contraction around 1% earths measurements. From stage 5's perspective, both the earths and stage 10's lengths will contract to about 80%. Stage 10 will view length measurements of other stages opposite that of earth with earths lengths at 1% and stage 5 at 80% that of stage 10's measurements.

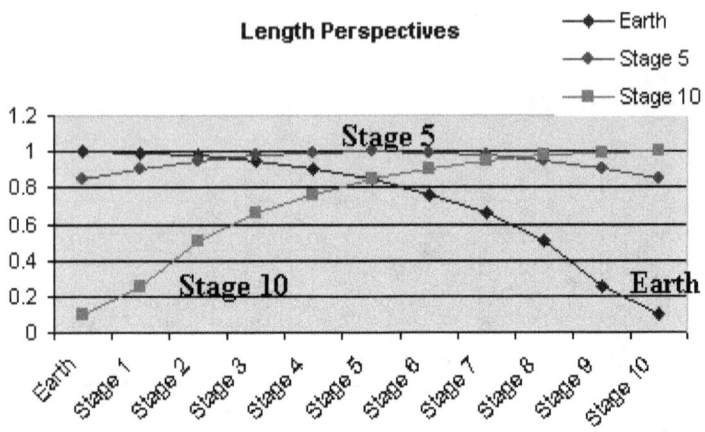

Length Perspectives

In another perspective, space would be compressed in front of a high-speed object similar to a bow wave in front of a boat and be stretched or dilated behind the object. Einstein had suggested that it was space – time that warped, not the moving objects.

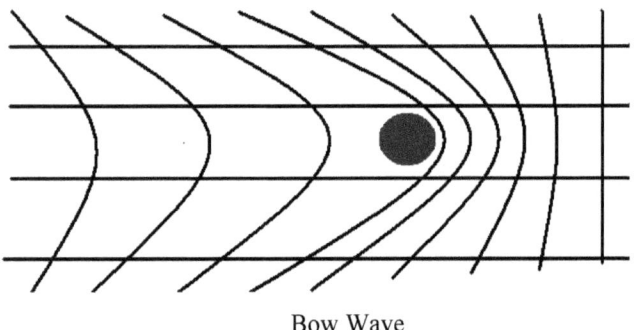

Bow Wave

If time is changed, the distance traveled must also change. Therefore, another perspective is to "warp" time in order to prevent Stage Two from going "Warp Speed." According to Einstein, the Dilation of Time would exponentially effect time the faster an object moves away or towards an observer.

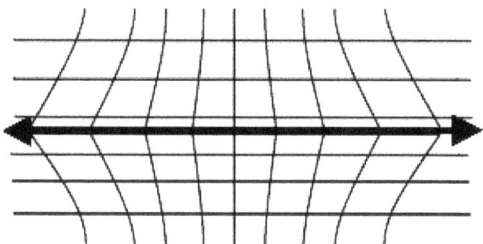

Dilation of Time

Going back to our rocket mind experiment, while viewing this rocketry from earth, Stage 5 hits terminal velocity at day 120, reaching a calculated velocity of 186,291 miles per second relative to the earth. At day 121, if it can continue to accelerate above light speed, it will reach a velocity of 188,509 miles per second, 8 million miles per hour faster than the speed of light. However, while viewing Stage 5 from Stage 1's observer's perspective, Stage 5 can accelerate on to day 146 where relative to Stage 1 it will hit terminal velocity at the speed of light. Remember that from Stage 1's perspective, he is stationary in space and the Earth is moving away from him in the other direction at 37,701 miles per second. The speed of light must have a reference; this reference will always be the observer. If there are two or more observers moving at different velocities in reference to each other and the same light source, the speed of light they detect will always be "c" to each of them.

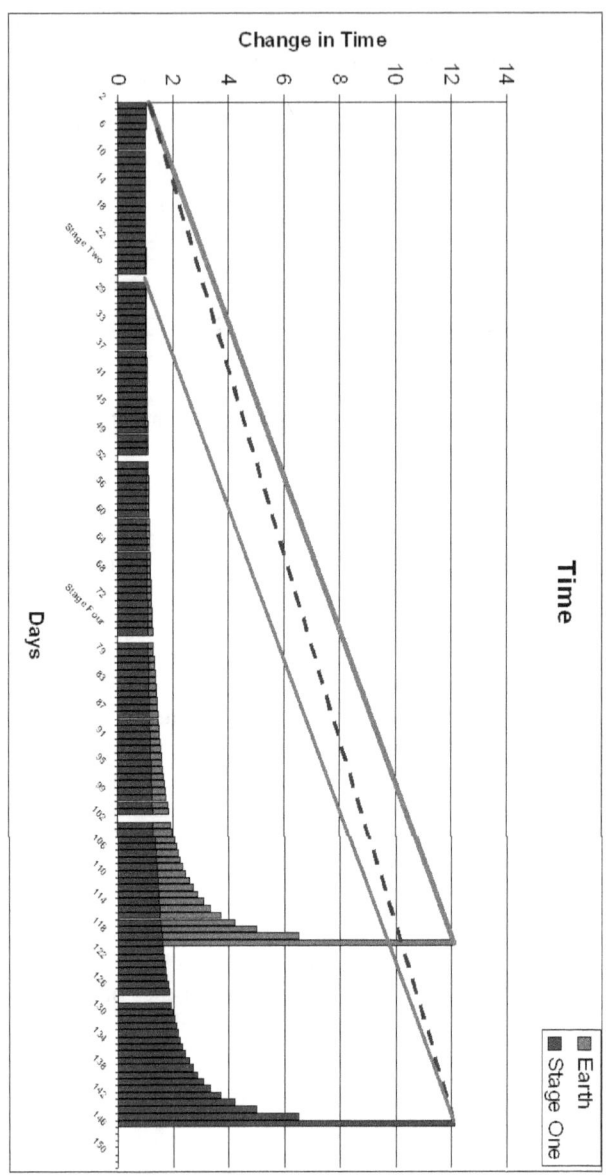

Time Shift Graph

In 1904 Henri Poincaré wrote:

"From all these results, if they were to be confirmed, would issue a wholly new mechanics which would be characterized above all by this fact, that there could be no velocity greater than that of light, any more than a temperature below that of absolute zero. For an observer, participating himself in a motion of translation of which he has no suspicion, no apparent velocity could surpass that of light, and this would be a contradiction, unless one recalls the fact that this observer does not use the same sort of timepiece as that used by a stationary observer, but rather a watch giving the "local time"

(Poincaré, 1904, p. 253)

Back to Basics

In the late 1890's, the pace of scientific discovery was colossal. Through the discovery of x-rays and other types of radiation, scientists such as **Henri Becquerel** as well as **Pierre and Marie Curie** began to put together the atomic model. Just after the turn of the century in 1911, **Ernest Rutherford** concluded through his research that a positive charge was confined to the small region at the center of the atom, what he called the nucleus; this was only fifteen years after the discovery of nuclear radiation by Becquerel and the Curie's (Bord, 2000, p. 441). Several years later, in 1913, Danish physicist **Niels Bohr** published an atomic model that better explained the natural atomic spectra using the now well-known "solar system" as a nucleus and electrons in distinct orbits around the atom (Bord, 2000, p. 381). The original model only subscribed to a positive and a negative charged particle model until in 1935, **James Chadwick**, working with **Irene Curie's** work (Marie Curie's daughter) discovered the neutron.

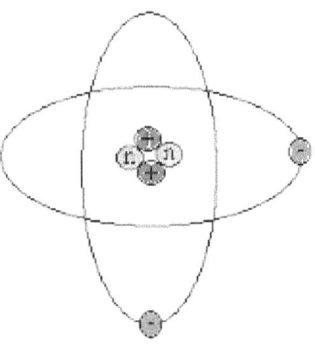

Bohr's Model of the Atom

The electron is the life of the atom. Electrons are believed to travel around the nucleus of the atom at fantastic speeds, some 700 miles per second and higher. Scientifically speaking, the electron is assigned a negative charge and the proton a positive charge, with opposite charges theoretically attracting each other equally. It is this high velocity in orbit that keeps the electron from falling into the nucleus of the atom as a result of the tremendous mutual attraction of the electrons and protons charges.

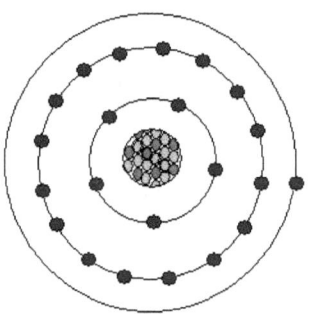

Conductor - One Electron in Outer Shell

Although the forces of the electrons charges are equal to that of the proton, the electron is approximately one trillionth of an inch in size with the proton around 1,800 times the electrons size. The neutron is similar in size to the proton but has no charge. A stable atom of any element normally carries a balance of multiples of these three atomic particles. A positively charged atom will carry more protons than electrons (a deficiency in electrons) and a negatively charged atom will carry more electrons than protons (Veatch, H., 1978, p. 22-23).

Today, about 155 combinations of these particles are known – called **elements**. Elements with a different number of **neutrons** than **protons** may have the same characteristics chemically but are considered an isotope and may be radioactive, behaving differently from a nuclear perspective (Bord, 2000, p. 413).

Some elements such as copper are called conductors and have the ability to share electrons in the outer orbits (shells) of their atoms. As a rule of thumb, elements with one, two or three electrons in their outer orbit are considered conductors, such as gold, silver and aluminum. These electrons move quite easily from atom to atom; this flow of electrons is called electricity. This sharing of electrons also produces the **Covalent Bond** that binds atoms together creating various chemical structures.

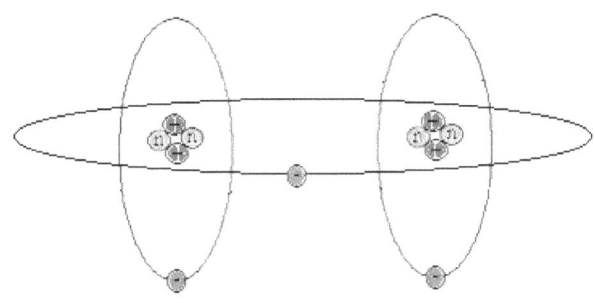

Conductor Sharing an Electron

Other elements have their outer orbits completely full and do not share electrons. These elements are considered insulators and resist electrical currents because their valence shell (outer shell) is full with valence electrons. The more electrons in these outer shells, the more inert the element is. Materials that fall in between insulators and conductors are called semiconductors such as carbon, silicon and germanium and are used primarily in electronic circuitry and integrated circuit chips (IC) used in electronics today (Evans, Alvis, p. 2).

Whether in chemistry or electronics, the electron is a source of energy that can be tapped. Electricity can be used to produce heat, magnetic energy in motors and electrochemical reactions such as chrome plating. Electricity is also used to create light and various types of electromagnetic propagation such as radio waves, microwaves, x-rays and gamma radiation. It is in understanding the electron that we can start to understand electromagnetic phenomena, energy and the fundamentals of light.

Energy

What is energy? Many eastern religions and even the new age movement speak of pure energy. Is there such a thing as "Pure Energy"?

Scientifically speaking, energy is a variable quantity that is a characteristic of an object. Energy is the ability to do work or that which is transferred when work is being done (Bord, 2000, p. G-1). There are several ways to describe energy, kinetic, potential, gravitational, electromagnetic, chemical and nuclear. Energy can be transformed from one form of energy to another such as light into heat, but cannot be created nor destroyed. This is known as the **Law of the Conservation of Energy** (Bord, 2000, p. 98) written in **Sir Isaac Newton's** "***Philosophiae Naturalis Principia Mathematica****,*" Latin for "Mathematical Principles of Natural Philosophy"- one of the most influential scientific works ever published.

Let's start with some energy formulas:

Velocity = Distance/Time. The rate of change in the position of an object over time.

Acceleration = Velocity/Seconds^2 Change in velocity of an object over time, over time.

Momentum = Mass x Velocity. The mass of an object multiplied by its velocity.

Inertia = A resistance of an object to the change in its state of motion.

As we can see, all of these formulas contain an object. Objects have mass, therefore "energy" simply put is a quantity associated with the movement of mass either kinetically (active) or potentially.

As a ball is thrown, it is accelerated to a specific velocity and has a certain momentum; the ball then decelerates and falls to the earth. The energy put into the ball dissipates through friction in the air and gravity. The remaining energy the ball has then is transmitted into and dissipated through its contact with the ground as vibrations and as sound through the air.

A rocket shot into space will remain at the velocity it was accelerated to once the engines are shut down. It will stay at this velocity and direction of travel until acted upon by another force such as gravity or friction due to reentry into the atmosphere.

Newton's First Law of Motion

"Every object in a state of uniform motion tends to remain in that state of motion unless an external force is applied to it"

(Bord, 2000, p.47)

Einstein's Most Famous Equation

As we look at Einstein's most famous equation we also see that mass has to be part of the energy equation.

$$E = MC^2$$

Here E is energy, M is mass and C is the speed of light at approximately 300,000 kilometers per second or 186,000 miles/second - identical to momentum - **Mass x Velocity** (Born, Max p. 283-286).

Subatomic Particles

There is a difference in how subatomic or elementary particles react with other particles and how they dissipate their energy. Unlike our world, where energy such as sound is dissipated through many different *medium*s and distributed into billions of particles, energy from subatomic particles can only be distributed and dispersed into fixed packets or bundles of energy that cannot be divided (Polkinghorne J. C., 2000, p. 6).

In the twentieth century, studies revealed that light could cause atoms to emit electrons. In 1900, while studying **Black Body Radiation** (BBR), **Max Planck** (1853-1947) found that atoms emitted energy in multiples of a certain energy unit, E or 1, 2, 3… and so on, multiples of E; it was not possible to emit a fraction of a unit (Polkinghorne J. C., 2000, p. 5). Einstein suggested that light energy may not be evenly spread across a given area but concentrated in small "packets" or "quanta". The energy transferred from one particle to another had to be completely transferred, all or nothing.

This localized energy was coined the "photon" (Kafatos, M. 1990, Pg 29). The "aether theory" was thrown out and the "corpuscle theory" also known as the "particle theory of light," originally suggested by Newton, was then reintroduced. The "aether theory" had too many questions that could not be answered. The "particle theory" solved many of these questions and then became the preferred theory of the scientific community.

Subatomic Expressions of Energy

Subatomic particles such as electrons demonstrate the energy they obtain in several ways.

Spin – Where the particle has angular momentum, a conserved quantity based on its mass and speed that will continue unless acted upon by another force (Polkinghorne J. C., 2000, p. 96). The spins of all known particles consist of a half spin or whole spin or a multiple of that factor, for example 0, ½, 1, 3/2 … etc. This energy is transferred from one particle to another in fixed quantities in particle-like packets of energy called quanta (Bord, 2000, p. 458).

Velocity – Where a particle has orbital angular momentum in orbit around an atom. This energy is also transferred from one particle to another in fixed quantities in particle-like packets of energy called quanta. As an electron absorbs a "quanta" of energy, its velocity and energy levels change causing it to "jump" from a low orbit to a higher orbit. Likewise, as an electron emits or looses a "quanta" of energy, its velocity and energy level again changes causing it to "jump" from a high orbit to a lower orbit (Bord, 2000, p. 381).

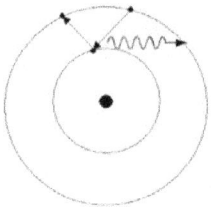

Electron Releasing a Packet of Energy

Frequency – The rate at which a particle oscillates per second (Polkinghorne J. C., 2000, p. 213). This frequency is proportional to the wavelength of the energy it disperses. This wavelength is the measured distance of a cycle from beginning to end. This electromagnetic energy wave travels at the speed of light (29,979,245,800 centimeters per second) and their frequency and wavelength can be determined by these formulas:

Frequency = $\dfrac{\text{Speed of light}}{\text{Wavelength}}$

Wavelength = $\dfrac{\text{Speed of light}}{\text{Frequency}}$

Speed of light = λ x Frequency   Wavelength = λ

The amount of energy a particle such as a photon has depends on the frequency at which it resonates. $E = hf$ where E is energy, h is Planck's constant at 4.136×10^{-15} Electron Volts (eV) and f is the frequency that the particle oscillates at in Hertz (Bord, 2000, p.375).

Examples:

Red light: f = 4.3 X 10^{14} HZ (Hertz) at **1.78 eV**

Blue light: f = 6.3 X 10^{14} HZ at **2.61 eV**

X –ray: f = 5 X 10^{18} HZ at **20,700 eV**

An electron in orbit is balanced between electrical attractive forces that pull it towards the nucleus while its centripetal acceleration caused by angular momentum keeps the electron from falling towards the nucleus. In the lowest orbit, orbit one; an electron has the smallest amount of energy possible. As it gains energy the electron moves out to larger orbits. Each "jump" to another orbit is caused by a gain or loss of packets of energy called "quanta" – Planck's constant. Each jump called a "radiative transition" causes the electron to radiate or absorb energy and change orbit. The larger the transition or "jump", the higher the frequency of the energy radiated. For example, the transition from orbit 3 to orbit 2 may give off the lower frequency red light burst whereas a "jump" from orbit 6 to 2 may give off a high frequency violet light burst. Electromagnetic spectroscopy analysis of these light burst show patterns that are unique to specific elements and can be used as fingerprints in determining the element present by the light they radiate.

The speed of an electron, as calculated by French physicist **Louis de Broglie** (1892-1987), in the smallest orbit of hydrogen atom is about $2.19*10^6$ m/s or meters per second (Asimov, Isaac, p. 86). The speed of light is $2.998*10^8$ m/s or 299,800 kilometers/sec (186,000 miles/sec). The mass of an electron is approximately $9.11*10^{-31}$ Kg. Therefore the momentum of an electron in the smallest orbit at $2.19*10^6$ m/s is $1.995*10^{-24}$ kilogram-meters per second (Bord, p. 385).

If this momentum is transitioned to the speed of light though the photon, the mass of the photon would be the energy or momentum of the electron divided by the velocity of the photon which is the speed of light or $(1.995*10^{-24})/(2.9979*10^8) = 6.655*10^{-33}$ Kg (compared to the electrons $9.11*10^{-31}$ Kg). This seems to suggest that the photon would be 136.89 times smaller in mass than an electron and 251,532 smaller than the mass of a proton. Likewise, if a photon's mass is 6.65496 $*10^{-33}$ Kg. and its velocity is the speed of light at $2.998*10^8$

m/s then = $1.995*10^{-24}$ kilogram-meters per second – enough momentum to cause an electron to "jump" from one shell to another in an orbit around an atom or the affect of "h", a Planck constant or "quanta".

The equation $E = MC^2$ indicates that the energy of a photon depends on its mass multiplied by its velocity (the speed of light) or E (photon) = mv^2. In reference to the speed of light this would be E = c(mv) or E = cP where (mv) or P is momentum.

There is a problem though; according to Einstein, as a particle increases in velocity closer to the speed of light, its mass increases exponentially – called relativistic mass compared to mass at rest. For example if an object whose mass is equal to 500 kg approaches the speed of light, say 90% the speed of light, its mass would increase to approximately 1143.8 kg - more than double its mass.

$$m = \frac{m_0}{\sqrt{1 - \frac{v^2}{c^2}}}$$

Relativistic Mass Formula

According to physicist, Satyendra Nath Bose, the photon must be different from other subatomic particles in that it must have zero mass at rest in order to go the speed of light (Isaacson, Walter p. 328). As the velocity (here referenced at c, the speed of light) of a particle decreases, the mass of the particle also must decrease to the point where it has no mass at rest (Ditchburn, R.W., p. 343).

All these fifty years of conscious brooding have
brought me no nearer to the answer to the question,
'What are light quanta?'

Albert Einstein   (Isaacson, Walter, p.101)

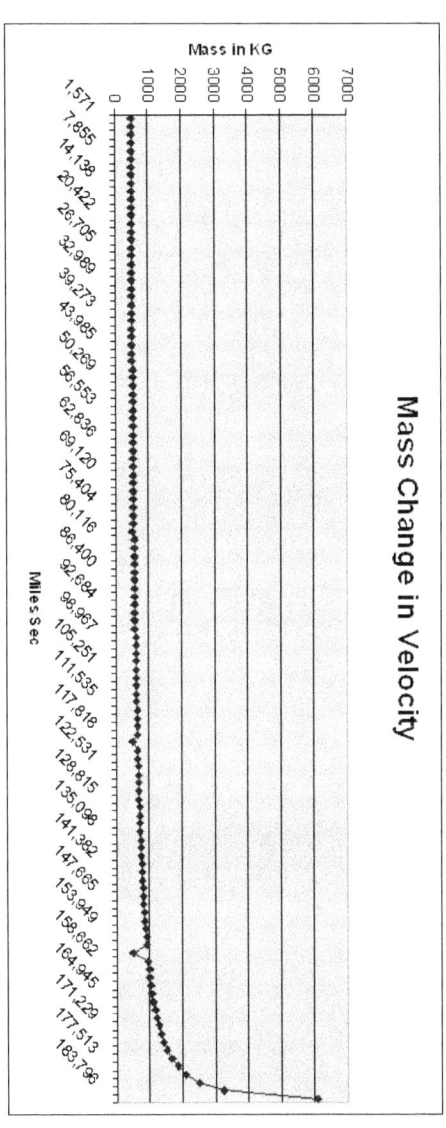

Mass Change In Velocty Graph

95

Review of Literature

Mankind has always wondered about the universe. Where did we come from and what are we made of? The questions about light, its composition and how it worked has always intrigued us, even back to the 300 BCE's, in the days of Plato, Aristotle and Socrates. In those days, light was considered the fifth element; the "luminiferous aether", the stuff stars were made of. It wasn't until Galileo in the early 1600's that the questions of lights speed was tested with the use of covered lanterns. All that he could determine at that time was that the speed of light was at least 10 times faster than the speed of sound. Not too much later in the late 1600's however, Ole Rømer determined through the study of Jupiter's moons that the speed of light was be 299,792 km/sec or 186,290 miles/sec, becoming the cornerstone of quantum physics, the study of atomic matter.

Various theories of what light consisted of, slowly developed, including René Descartes undulation theory, otherwise known as wave theory in the early 1600's and Sir Isaac Newton's corpuscular theory or particle theory in the early 1700's. Newton's particle theory however did not stand very long as Thomas Young in the late 1700's used the "Double Slit" experiment where light created interference patterns, proving that light was composed of waves. Studies showed that light was made of both longitudinal waves and transverse waves complicating the aether theory of the day. This suggested that the aether, the medium that light traveled through, must be created of a solid substance, which would resist the motions of the planets and the stars and did not seem to make sense.

In the 1700's, James Bradley became the first to prove the Copernican heliocentric view of the universe. It became evident through his studies when viewing stars through his telescope that the earth moved very rapidly though space in its orbit around the sun. These studies of star aberration, or the apparent shift in a stars position, showed different displacements of the stars over time in the northern region around the North Star and south in the equatorial region. This lead to his calculation of the speed of light which was very close to Ole Rømer's calculations 100 years earlier. This knowledge was reinforced by Fizeau's and Foucault's experiments showing lights speed within a couple thousand miles per second.

Later on in the late 1700's François Arago tried to show that light traveled at different speeds due to gravitational forces of stars and the movement of the stars. His experiment, using a prism and a telescope was set up to detect these various speeds using Newton's refraction theory. The test failed; at least in showing that there was no variation in the refraction of light. This experiment added confusion to the mix by indicating that the speed of light from large stars and small stars, and that of moving stars in relation to the movement of the earth was the all same.

The aether or medium, through which light moved, at this time, was thought to be dragged along with the earth through space. Star aberration however disproved this, showing that the aberration was due to lights displacement by the earth's movement and was not dragged along by the atmosphere. French physicist Augustin Fresnel would then develop the partial aether drag theory, suggesting that light traveling through air would not be dragged along, but that as the refraction index of a substance increased, so also would the drag effect increase. Light would therefore travel unimpeded through the air and be uninfluenced by the movement of the earth. Once light enters into glass or some other medium of higher refractive indexes it would therefore be constrained or locked into the new medium and travel at one velocity with that medium.

George Biddell Airy later in the 1800's tried to detect the aether using a kind of water filled telescope. The theory was that light traveled slower through water and that while observing star aberration due to earth's movement through space, the aberration angle would change. This also failed; indicating that the aether or medium though which light traveled did not exist, adding perplexity to this study. In the late 1800's Albert Michelson and Edward Morley went on to disprove the existence of the aether with their interferometer experiment.

Willem de Sitter, early in the 1900's discovered that light traveled the same speed in his studies of twin stars; much like Arago's experiments in the 1700's. His discoveries showed the Doppler Effect evident in the shifting of the light spectrum towards the red or blue due to the movement of these stars. Although the speed of light remained the same to the observer, the frequency of the light viewed changed proportional to the speed and direction the star was moving in reference to the observer.

These experiments confirmed to scientists of that day that there was no aether. Not only did these tests show that there was no aether, they had also discovered what is now called "Law of the Constancy of the Speed of Light." This measurement of the speed of light is the same no matter what the observer's velocity is what velocity the emitter is, either moving towards or away from each other expressed as "c" and is measured at 299,853 km/sec. or 186,328 miles/sec.

Hedrick Lorentz would then try to rectify this strange problem by developing the "Lorentz Contraction" theory. This theory showed that the movement of an object causes length contraction of that object in the direction of its movement, therefore causing the speed of light to remain constant no mater what the movement of the observer or object being observed. This is one explanation in how the Muon can reach the earth at high speeds even though its life cycle supposedly would prevent it. The "Lorentz Contraction" due to its velocity at near light speed, changes the space-time environment of the Muon, allowing it to reach the ground.

Albert Einstein would then build on this to develop his "Theory of Relativity". Einstein theorized that movement changed time and space and that the faster an object moves the more time and space around it would change. Gravity also would have a dramatic effect on space and time causing light waves to bend around stars and for time to vary in different gravitational fields. This explains the constancy of the speed of light and the shift in the light spectrum due to gravitational fields and light from objects at different velocities.

Modern day studies seem to verify Einstein's theories beginning with studies on the precession of Mercury. Studies of starlight being bent by the suns gravity, gravitational lensing due to the intense gravitational fields of galaxies also validates this theory. Galactic redshift and gravitational redshift is seen in the GPS transmission or radio waves and even the tracking of Venus and Mercury with radar. It seems that no matter how you look at it, everything seems to indicate that Einstein was correct.

Edwin Hubble's discovery in 1929 showed that while studying galactic red shift that redshift in light from stars and galaxies increase in proportion to their distance from earth. This developed into what is now known as Hubble's Law, expressed by the equation $v = H_0D$. The further we look out into the universe, the faster galaxies and stars accelerate away from each other, indicating that these galaxies and stars were moving out from a single point in space over a long period of time.

This discovery lead to the "Big Bang Theory" as scientist began to work back in time to a point where this rapidly expanding universe theoretically began, to a one point in space and time called the "singularity" over 13 billion years ago. In 1964, Arno Penzias and Robert Wilson would discover Cosmic Microwave Background Radiation (CMBR), evidently a relic of the "Big Bang" dispersed evenly across the sky.

Studies from observing these stars and galaxies generated new ideas, one being Dark Energy. What was the force that acted as a dispersing agent, pushing the galaxies and stars away from each other in deep space; an energy that was repelling against the forces of gravity, forces that should pull everything together in what eventually would be the "Big Crunch"? Another recent observation also suggests the existence of Dark Matter, an invisible mass that would generate enough gravity to keep the rapidly spinning galaxies from flying apart. Existing "known" mass in galaxies such as our own Milky Way, does not have the amount of mass needed to keep stars on its outer edge, moving at great velocities, from flying off into space.

Review of Theory

Einstein and the other scientist at the turn of the 19th century were baffled by the fact that the speed of light was always the same velocity to an observer, no matter what the velocity of the observer or the velocity of the object emitting the light. Lorentz had shown that if an object was moving, that it contracted in the direction of its motion, increasing the distance light had to travel, therefore allowing the "perceived "speed of light to remain the same. Motion was warping space and time. This led Einstein to develop his theories of Relativity.

The Theory of Special Relativity explains that motion is relative; Einstein showed this with the train and embankment thought experiment. Here two bolts of lightning strike simultaneously in front and behind a moving train. There is one observer midpoint between the lightning strikes on the moving train and one observer midpoint the lightning strikes on the embankment beside the train. Although the lightning strikes occurred at exactly the same time to the observer on the ground, the observer on the moving train saw the light from the front of the train first because of his forward motion. By the time the light meets this observer, he has moved towards the lighting in front of the train and away from the lightning behind the train.

In this text, we have reviewed a similar thought experiment with a train blowing its whistles simultaneously from the engine and caboose. We saw the different times the whistle was heard by the two observers and also saw the Doppler Effect by the movement of the train. Here, movement or motion caused a different reality for both observers.

The Theory of General Relativity went on to show that mass also has an effect on time and space. Mass causes space to bend around it, which in turn causes mass to attract mass, the effects of gravity. This also is what causes light waves to bend and slow down, manifesting itself in gravitational lensing and red shift. The effects of mass also cause clocks to change their tempo, in effect changing time. Therefore both mass and velocity have a direct influence on time and space. The greater the mass, the more space and time are warped and the greater the velocity, the greater time and space are warped.

101

Einstein also showed that gravity and acceleration are equivalent with his Equivalence Principle. This was shown with a thought experiment where an observer was floating in space in a large room with no gravity. A force was applied to the room that caused it to accelerate at a constant rate, forcing the observer onto the floor. The observer, not being able to see outside, assumed that this force was gravity. This asserted that gravity and acceleration is essentially the same thing.

Another thought experiment was performed in this text in Blast Off. Here velocity was studied comparing the speed of one rocket stage to another in outer space, showing that velocity was relative to a reference point – one of the other stages. This point was also shown with the baseball being thrown from a moving vehicle earlier in this text. The point here given was to show that if speed was relative, the effects of speed would also have to be relative. This means that since velocity changes time, that this time change would be relative to different observers at different relative speeds.

As mass increases at speeds close to the speed of light, this mass change must also be relative to different observers. An object would therefore be getting larger to one observer than to another observer at a different relative velocity; how can this be? Just as in Einstein's train, this can only be a **perceived** difference and not reality. The only way that we can get to a point where the effects of velocity are real is to create a stationary point in space that all velocities are referenced to.

Conclusion

Even as far back Aristotle and Plato, mankind has always been curious as to what light truly is. It was Galileo who first verified that the speed of light must have been at least ten times the speed of sound with his lantern experiment in 1638.

From then on, experiment after experiment, such as Roemer's studies of Io's movement around Jupiter, proved that the speed of light was finite and eventually believed to be "absolute", now known as "c". No matter how the scientists looked at light, however, there seemed to be a conflict, light seemed to have a dual personality, both a particle and a wave – known as the Wave - Particle Duality. Thomas Young's experiments proved that light was made up of waves as far back as the late 1700's. Max Plank showed in 1900 that light was made up of fixed quantities of energy he called "quanta" showing light to be made of particles now known as photons.

Einstein explained this as the "photoelectric effect", proposing the light traveled in fixed quantities, like bundles, now explained as the Photon Theory. The Michelson- Morley experiment had proven that there was no aether as previously believed. Even today there still seems to be a conflict; experiments still show that light functions both as a wave and a particle, depending on how the experiment is set up and observed. This speed of light however is now considered to be constant leading to Albert Einstein's Theory of Relativity and the mathematical equation, $E=MC^2$.

Appendix

I cannot but regard the aether, which can be the seat of an electromagnetic field with its energy and its vibrations, as endowed with a certain degree of substantiality, however different it may be from all ordinary matter.

(Lorentz, 1906)

(Borel, Armand, p. 690)

Indeed one of the most important of our fundamental aether not only occupies all space between molecules, atoms, or electrons, but that it pervades all these particles. We shall add the hypothesis that, though the particles may move, the aether always remains at rest.

(Lorentz, 1906)

(Lorentz, H.A, p. 11)

Lorentz proclaimed the very radical thesis, which had never before been asserted with such definiteness: the aether at rest in absolute space.

In principle this identifies the aether with absolute space. Absolute space is no vacuum, but something with definite properties whose state is described with the help of two directed quantities, the electrical field E and the magnetic field H, and as such, it is called the ether.

(Born, 1924)

(Born, Max, p. 204)

Index

References

Arago, François (1858) Oeuvres Complètes, 7th Edition, Volume 4,
    Paris

Aristotle (350 B.C.E) De Caelo, Book I, Chapter 2

Asimov, Isaac (1991) Atom, Journey Across the Subatomic Cosmos,
    Truman Talley Books, First Plume Printing, New York, NY

Barbour, Julian B. (2001) End of Time: The Next Revolution in
    Physics, Oxford University Press, 198 Madison Avenue New
    York, NY

Bartusiak, Marcia (2000) Einstein's Unfinished Symphony, Listening
    to the Sounds of Spac^{2}e -Time, Burkley Publishing Group, New
    York, NY

Bless, Robert C. (1996) Discovering the Cosmos, University Science
    Books, 55D Five Gates Rd, Sausalito CA

Bord, Donald J. (2000) Inquiry Into Physics, Brooks / Cole Fourth
    Edition.

Borel, Armand (2001) Oeuvres, Collected Papers, Vol. IV, Springer
    Scientific and Business Media, 1150 65th Street, Emeryville, CA

Born, Max, (1962) Einstein's Theory of Relativity, Dover
    Publications Inc. New York, NY

Brown, Julian, (2000) Minds, Machines, and the Multiverse, The
    Quest for the Quantum Computer, Simon and Schuster, New
    York, N.Y.

Collins, Harry (1998) The Golem: What You Should Know about
    Science, Cambridge University Press, Cambridge CB2 2RU, UK

Dekel, Avishai (1999) Formation of Structure in the Universe,
    Cambridge University Press, Cambridge CB2 2RU, UK

Dennis, Mark F. (2009) A Challenge to Einstein, Golden Iris
    Publications

Ditchburn, R.W. (1991) Light, Dover Publications, Inc., 31 E.2nd
    Street, Mineola, NY

Einstein, Albert, (1952) Relativity by Einstein, Special and General
    Theory, Tess Press, Fifth Edition

Evans, Alvis, J. (1985) Basic Electronics Technology, Tandy Corporation, Ft. Worth, Texas

Fresnel, A. (1818) Lettre d'Augustin Fresnel à François Arago Annales de chimie et de physique. 9, p. 57–66

Galileo, Galilei (1632) Dialogue Concerning the Two Chief World Systems, Second Edition, University of California Press, Berkeley, CA

Glendenning, Norman K. (2007) Our Place in the Universe, World Scientific Publishing, 27 Warren Street, Hackensack, NJ

Hawking, Stephen (1988) A Brief History of Time, Bantam Books, Random House Inc. New York, N.Y.

Hawking, Stephen (1993) Hawking on the Big Bang and Black Holes, World Scientific Publishing, P.O. Box 128, Farrer Rd., Singapore

Hobson, Michael P. (2006) General relativity: An Introduction for Physicists, Cambridge University Press, Cambridge University Press, Cambridge CB2 2RU, UK

Isaacson, Walter (2007) Einstein, His Life and Universe, Simon and Schuster Paperbacks, New York, NY

Kafatos, Menas (1990) The Conscious Universe, Springer-Verlag, New York, NY

Lorentz, H.A, (1909) The Theory of Electrons, B.G. Teubner, G. E. Stechert, Booksellers, New York, 129- 133 West 20th St. New York, NY

Newton, Sir Isaac, (1730) Opticks, a Treatise of the Reflections, Refractions, Inflections and Colors of Light, Fourth Edition.

Poincaré, H. (1904) The Present and the Future of Mathematical Physics, Address Delivered to the International Congress of Arts and Science, St Louis.

Polhinghorne, J.C. (1984) The Quantum World, Princeton Science Library, Princeton University Press, Princeton NJ

Sokolov G., (1999) The Theory of Relativity and Physical Reality

Veatch, Henry (1978) Electric Circuit Action, Science Research Associates, Chicago Ill.

Van der Kamp, Walter (1988) De Labore Solis, Airy's Failure Reconsidered, Anchor Book & Printing Centre, Surrey, B.C., Canada

Wolf, Fred, (1989), Taking the Quantum Leap, The New Physics for the Non-Scientists, Harper and Row Publishers, New York, NY